AF608302

ON THE MARGINS OF URBAN SOUTH KOREA

Core Location as Method and Praxis

On the Margins of Urban South Korea

Core Location as Method and Praxis

EDITED BY JESOOK SONG
AND LAAM HAE

UNIVERSITY OF TORONTO PRESS
Toronto Buffalo London

Toronto Buffalo London
utorontopress.com

ISBN 978-1-4875-0335-2

Library and Archives Canada Cataloguing in Publication

Title: On the margins of urban South Korea : core location as method and praxis / edited by Jesook Song and Laam Hae.
Names: Song, Jesook, 1969– editor. | Hae, Laam, editor.
Description: Includes bibliographical references and index.
Identifiers: Canadiana 20190153768 | ISBN 9781487503352 (hardcover)
Subjects: LCSH: Community development, Urban – Korea (South) | LCSH: Sociology, Urban – Korea (South) | LCSH: Decolonization – Korea (South) | LCSH: Knowledge, Sociology of.
Classification: LCC HT169.K6 O5 2019 | DDC 307.1/416095195—dc23

The mural featured on the front cover of this volume is one of many painted in Ihwa-dong in Seoul, South Korea. The neighbourhood is also called Ihwa-dong Mural Village (Ihwa-dong byeokhwa maeul). The particular mural featured on the front cover memorializes textile workers (mostly underpaid female workers) who used to work in sweatshops located in Ihwa-dong from the 1970s to the early 1990s during the industrialization period in South Korea. From the mid-2000s, murals have been painted in Ihwa-dong as part of the central and local governments' urban beautification project for dilapidated neighbourhoods. The increasing popularity of the murals in Ihwa-dong and the subsequent touristification of the neighbourhood has caused a wave of controversies and tensions among local residents. The editors of this volume decided to put the mural of female textile work on the front cover to point to the irony of historical memory and visual memorialization and urban redevelopment, which echoes the themes developed in many of the chapters in this volume. Photograph by Laam Hae.

University of Toronto Press acknowledges the financial assistance to its publishing program of the Canada Council for the Arts and the Ontario Arts Council, an agency of the Government of Ontario.

Canada Council for the Arts
Conseil des Arts du Canada

Funded by the Government of Canada
Financé par le gouvernement du Canada

Contents

Acknowledgments

This volume would not have been possible without support from many people and institutions. The contributors to the volume deserve particular recognition and our gratitude: Hyeseon Jeong, Mun Young Cho, Seo Young Park, Sujin Eom and Youjeong Oh (alphabetical order for first names here and throughout this volume). We would like to thank the discussants and reviewers who provided invaluable comments on various iterations of the manuscript at conference and workshops – Albert Park, Bae-Gyuoon Park, Eleana Kim, Hyun Bang Shin, Jim Glassman, and Katharine Rankin – and two anonymous reviewers who read the full manuscript for the University of Toronto Press: all of these comments strengthened the manuscript. We are also grateful to participants in the workshops at Seoul National University, York University, and the University of Toronto, and to graduate students who took courses we offered in related subjects of Asia as method and postcolonialism. Although we do not list their individual names here, they were valuable interlocutors who helped us develop our thinking that is reflected in this volume.

We are also grateful to support from Kyle Gibson, who helped edit our manuscript and many times provided valuable input based on his specialty, and Grayson Lee for multiple rounds of reviewing bibliographies. Further, we appreciate the contribution of the romanization experts we consulted on the new style we decided to follow both in North America and South Korea. We owe big thanks for the many efficient and capable administrators who managed numerous hectic situations and logistics related to hosting events in Toronto: Alicia Filipowich, the coordinator of York Centre for Asian Research (YCAR) at York University; Annette Chan, the financial officer of the Department of Anthropology, University of Toronto; Grayson Lee, the student assistant at the Centre for the Study of Korea (CSK) at the University of Toronto; Minyoung Kye, at the Korean Office for Research and Education (KORE); and Rachel Ostep and Eileen Lam, then key administrators for the Asian Institute (AI) at the University of Toronto. Our appreciation also goes to directors of sponsoring institutions and programs, including Edward Banning, then the chair of the Anthropology Department at the University of Toronto; Jennifer Chun and Yoonkyung Lee, consecutive heads of CSK; Hyun Ok Park, the director of KORE; Philip Kelly and Abidin Kusno, consecutive directors of YCAR; and Rachel Silvey, the director of AI. These institutions, along with the Office of the VPRI and the Faculty of Liberal Arts and Professional Studies at York University, provided crucial support, such as work spaces and hours of labour by key administrators, and financial support for inviting people to workshops and/or subvention of book production. We would like to acknowledge that our volume was also supported by the Core University Program for Korean Studies through the Ministry of Education of the Republic of Korea and the Korean Studies Promotion Service of the Academy of Korean Studies (AKS-2018-OLU-2250001), a Social Sciences and Humanities Research Council Insight Grant (SSHRC 435-2016-0137), and a Social Sciences and Humanities Research Council Institutional Grant (Anthropology, University of Toronto, 2016).

We would also like to acknowledge all the support and hours of works by the University of Toronto Press, including Doug Hildebrand and Jodi Lewchuk, our acquisition editors; Barbara Tessman, our copy editor; and Robin Studniberg, production manager. It was pleasure to work with each.

To our families, friends, and care providers, we also want to express our deep gratitude. Without their support, patience, and care, we would not have found the time and energy to produce this volume.

Editors' Note

Unless indicated differently, Korea implied as a nation-state government or society refers to South Korea in this manuscript. We would like to note that we have followed the National Institute of Korean Language's guide for the romanization system for the English transliteration of Korean, which was established on July in 2000 by the Ministry of Culture Sports and Tourism. Also, we follow the authors' preference in their mother tongue tradition for the spelling and order of names of Baik Young-seo and, later, Sun Ge (Baik and Sun are last names). For authors whose works are known to anglophone readers, such as Kuan-Hsing Chen, we follow the order and transliteration used in English publications.

Abbreviations

ACHR	Asian Coalition for Housing Rights
ACPO	Asian Committees for People's Organization
BHA	Branksome Hall Asia
CO	community organizing
CRGP	Committee for the Revision of the Greenbelt Policy
EWP	Education Welfare Project
FRSN	Four Regions Slum Network
IMF	International Monetary Fund
JDC	Jeju Free International City Development Center
JFIC	Jeju Free International City
JGEC	Jeju Global Education City
JSSGP	Jeju Special Self-Governing Province
KACO	Korean Action for Overseas Community Organization
KONET	Korean Community Organization Network
KRIHS	Korea Research Institute for Human Settlements
LOCOA	Leaders and Organizers of Community Organization in Asia
MOCT	Ministry of Construction and Transportation
MOE	Ministry of Environment
MOLIT	Ministry of Land, Infrastructure and Transport
MOLTMA	Ministry of Land, Transport and Maritime Affairs
NGO	non-governmental organization
NLCS	North London Collegiate School
ODA	official development assistance
OECD	Organization for Economic Co-operation and Development
PASG	People's Action for Saving Greenbelts
PMG	Pohang municipal government
PRC	People's Republic of China
PSA	Pohang Saemaul Association

PV	Peace Village Social Welfare Corporation
SMR	Seoul Metropolitan Region
TCPA	Town and County Planning Association (Britain)
UPC	Urban Poor Consortium

ON THE MARGINS OF URBAN SOUTH KOREA

Core Location as Method and Praxis

Introduction: Core Location, Asia as Method, and a Relational Understanding of Places

LAAM HAE AND JESOOK SONG

This volume seeks to provide rich and illuminating accounts of the peripheries of urban, regional, and transnational development in South Korea. It is the outcome of long-term and ongoing interdisciplinary collaborations and dialogues among scholars based in a variety of disciplines, including architecture, anthropology, geography, and political science. The key threads that bind each chapter together are the ideas of "core location" (*haeksim hyeonjang*), a term coined by Baik Young-seo (2013a, 2013b), and "Asia as method," a concept with a century-old intellectual lineage in East Asia, especially as developed by Kuan-Hsing Chen. Each chapter offers an empirical account of different sites in Korea. The focus on sites may sound counter-intuitive in light of current trends toward conducting transnational studies in the social sciences, especially in the field of area studies. While our focus is on individual sites, however, our optic is not localist; rather, our approach is a *relational* one, situating individual sites within the broader matrix of social changes occurring at the urban, national, regional, and global scale. A "site" is an interconnected place where different forces and processes intersect and often contradict one another to produce and constitute a particular constellation. In this volume, we examine the constitution of different sites in Korea and aim to understand these interconnections, especially through the frames of core location and Asia as method. These conceptual apparatuses, which are rooted in a long intellectual tradition in East Asia, proffer a *reflexive* perspective, compelling us to re-examine inherited and taken-for-granted categories and theories, and enabling us to embark on the decolonization of our research. Furthermore, they compel us, as academics, to bear in mind the issue of praxis – of theoretically informed political action. Accordingly, examining oppositional politics within different places, and analytically and politically linking these places with variegated oppressive and exploitative systems, is a key mandate of each chapter in this volume.

Despite being little known in anglophone scholarship, Baik's concept of core location has gained some currency among East Asian scholars, especially among those affiliated with the journal *Inter-Asia Cultural Studies*. Initially introduced as a heuristic device to understand geopolitical conditions in East Asia, "core location" refers to a place with lived experiences of multiple layers of marginality. It is, however, not only about a particular geographical site of marginalization. Core location, for Baik, is a prism through which to capture and problematize multiple, contradictory, and convoluted layers of power stemming from colonialism, imperialism, militarism, and Cold War and post–Cold War dynamics that characterize the particular geohistory of East Asia and that are deeply entrenched in people's lifeworlds at particular locations in East Asia. In particular, Baik attends to the dynamics of power struggles between transpacific imperial powers, such as China, Japan, Russia, and the United States, as forces that have shaped the marginal states of core locations. His prime example of a core location is Okinawa. Having been annexed by Japan in 1879 and occupied as a US military base since the end of the Second World War, Okinawa has been a site in which Japanese colonialism, US military imperialism, and sexual violence by American servicemen have become embattled issues.

The concept of core location is not concerned only with understanding and interpreting a particular location and struggles projected through it; it also explores what forms of praxis can emanate from this understanding. Rather than simply trying to understand a core location as a victim of imperial power struggles, Baik argues that it is precisely within these core locations, sites of the downtrodden, that the potential to generate new politics and regional and global solidarity lies. For example, Okinawa has been the site through which solidarity movements across East Asia and Southeast Asia were spearheaded against Japan's past and current imperialist violence, its far-right nationalism, and US military imperialism and militarized violence in the Asia-Pacific region. In a similar vein, Baik points to the Korean peninsula as a core location that is fraught with contradictions stemming from Cold War and post–Cold War dynamics.[1] For him, the division between North and South Korea is the embodiment of the sort of ongoing Cold War politics that implicates both Euro-American imperialists and fascist factions in Japan (Baik 2013b, 157; see also Paik 2013). Pointing to a similar context to Okinawa, Baik emphasizes the importance of the development in Korea of anti-American peace movements

1 Another example that Baik elaborates on is Taiwan's internal colony over aboriginal people (Baik 2013a).

and activism for reparations for Japanese war crimes, and he contends that these movements and activism are important fields for scholarly research. For Baik, the production of socially engaged knowledge is of the utmost importance. The three pillars of Baik's ideas are critical self-reflection (*seongchal*), praxis (*silcheon*), and communicative connection (*sotong*) (Baik 2012, 455). His concern is to discover and build a common ground, a universal base that connects different sites of resistance, but he argues that this universality should be based on profound insights about a place (*tongchalseong ui bopyeonseong*).

Kuan-Hsing Chen's "Asia as Method": Toward the De-imperialization of Knowledge

Baik's notion of core location not only intervenes at the level of ontology and politics in relation to East Asian spaces and scholarship. The concept is also loaded with a particular epistemology. According to Baik, it is at, and through, core locations that we can identify forms of "double marginality" (*ijungjeok jubyeonui sigak*) (Baik 2013a, 17–18, 45).[2] The first form of marginality refers to the people inhabiting downtrodden places who have been relegated to the margins within the geopolitical hierarchy within, across, and beyond Asia. The second form of marginality is the peripherialization of place-rooted standpoints in East Asia that have been rendered invisible under the hegemony of Western-centred world historiography and scholarship. Baik argues for a centring of the perspectives of East Asia away from this peripherialization. This problematic that Baik raises resonates with the long tradition of Asia as method, as was developed by Takeuchi Yoshimi and Mizoguchi Yūzō and, more recently (and often collaboratively), by Sun Ge and Baik Young-seo.[3] It especially echoes the key argument that Kuan-Hsing Chen expresses in *Asia as Method* (2010).[4]

---

2 Some chapters in this volume refer to the idea of "double marginality" (*ijungjeok jubyeonui sigak*) as "twofold-peripheral perspective" or "two-fold peripheries," following Baik's own expression in English (2013b, 145).

3 Each of these authors has intervened differently in the tradition of Asia as method. But what binds these different intellectuals together is their interest in self-reflexivity, critical perspectives on oneself and others, self-transformation by understanding others, and understanding the world through the perspectives and lives of people, especially those most marginalized via historical injustices of wars and imperialism (see Yoshimi [1960] 2005; Mizoguchi [1989] 1996); Sun 2003, 2007, 2013; Sun and Yoon 2013; Yoon 2014).

4 Chen has mobilized, and collaborated with, other East Asia–based intellectuals to found the journal *Inter-Asia Cultural Studies* and hold biannual conferences of the same name, which Baik also participates in.

In this work, Chen calls for the de-imperialization and de-Westernization of knowledge production. He argues that Western concepts that are premised on capitalist modernity "render everything else invisible or irrelevant" and therefore offer "inadequate analytical understandings of our own [Asian] societies" (2010, 224). Under Euro-American dominance, Asian history and historiography have become "a footnote that either validates or invalidates Western theoretical propositions" (226), and Western modernity and its theories become "the standard against which all other places are measured" (253). Chen urges that scholars challenge the process in which the West became the single reference point in the processes of knowledge production and circulation. For him, the particular geohistories of colonialism, imperialism, and Cold War and post–Cold War dynamics of Asia, as well as the liberalization and democratization processes in each country in Asia, reveal a different world history and historical perspective from the one in which the West has been central.

Taking issue with the practice of using the West as a reference point to understand other places, Chen highlights the urgency and importance of "multiplying and shifting our points of reference" (224). In particular, his interest lies in developing co-referencing between different countries in Asia, and he argues that Asians can come to grips with problems in their respective locations by inter-referencing with the structural problems and the resistant politics developed to combat them in each other societies, instead of looking toward the West for understanding and solutions (212). To this end, Chen engages with subaltern studies developed by Indian postcolonial scholars. In particular, Chen examines Partha Chatterjee's (2004) notion of "political society," a term that Chatterjee develops to explain the experience of Indian modernity, thus challenging the Western modern paradigm of "civil society" that does not entirely capture social formations in India. Here, Chen develops the method of inter-referencing to better explain Taiwanese society.[5]

Chen's Asia as method parallels the problematics raised by postcolonial studies. Postcolonial theories have stressed the world's heterogeneity, rejected historicism, and emphasized the local specificity and

---

5 Chen was influenced by the work of Mizoguchi Yūzō, the author of *China as Method*. Drawing upon Mizoguchi, Chen argues further that "Asia as method" is a project of transforming Asians, "a precondition for arriving at different understandings of the self, the Other and world history" (253). Sun Ge also notes that Yoshimi, inspired by Lu Xin, stresses this aspect to criticize Japanese Sinologists who condescendingly objectified China (Yoon 2014).

multi-linearity of historical progression (Chakrabarty 2000). Postcolonialists have also asserted that local historical developments have been "judged almost exclusively against a European norm, and those histories which did not fit or comply with that norm were dismissed as 'incomplete'" (Anievas and Nişancıoğlu 2017, 44). Marxism's renditions of universalism and specific teleology, in particular, have been the primary target of postcolonial critique. Critics have argued against the Marxist resort to such binaries as those of pre-capitalism and capitalism, premodern and modern, pre-political and political, which do not capture the totality of life in non-Western societies (ibid.). These binaries, as well as other Marxist theoretical constructs, are not universal; according to postcolonial thinkers such as Chakrabarty (2000), they are rooted in the particular history of Europe and are, therefore, provincial.

Postcolonial scholarship has also influenced various disciplines within the area studies field, and there have been initiatives among area studies scholars to rewrite the history of each specific region against a Western-influenced historiography that is often closely associated with particular claims of "scientific truth" and universalism. Yet this new type of area studies scholarship has often been subject to criticism because it reifies native cultures, over-emphasizes insiders' knowledge, and denies that "the West" is already internal to the consciousness of natives (Dirlik 2005, 163). These area studies as well as postcolonial works in general have also been criticized for rejecting any form of universality and dismissing the broader political economic structures that have continued to generate violence, dispossession, and exploitation in different parts of the world (Dirlik 1994; Chibber 2013). For scholars such as Dirlik (1994), Harvey (1989), and Jameson (1991), postcolonial studies prioritizes discursive aspects of power, and they argue that its emergence is an expressive ideology of late, post-Fordist capitalism.

Chen's approach is in certain respects more nuanced than the types of postcolonial studies that these critics have found fault with. For instance, he argues that Asia as method, like Chakrabarty's (2000) project of provincializing Europe, is not a nativist or atavistic project (Chen 2010, 219). Chakrabarty, while emphasizing that "getting beyond Eurocentric histories remains a shared problem" (2000, 17) among postcolonies, also asserts that "provincializing Europe is not a project of rejecting or discarding European thought" (16): European thought is "both *indispensable* and *inadequate* in helping us to think through the various life practices that constitute the political and the historical" in different locations (6, emphasis added). In this regard, we agree with Anievas and Nişancıoğlu's (2017) argument that Marxist criticisms of

Chakrabarty that stress his supposed denial of Western ideologies' presence in the East, especially Chibber's (2013), are based on a misreading of his views. Similar to Chakrabarty, Chen (2010) maintains that it is important to acknowledge that the West is already entangled in the East, and that the West exists "as bits and fragments that intervene in local social formations in a systematic, but never totalizing, way" (223). The West as fragments, in other words, becomes "internal to the local," "one cultural resource among many others," and is an inalienable, if partial, part of Asian subjectivity (223). Therefore, Asia as method is not a project that is concerned with a sort of Asian particularity that makes Asia incompatible with the West. The study of China as method, for example, does not represent a search for an essentialized, fundamental core that is the "real" China. Such reasoning is vulnerable to the political manipulation of orientalist ideologies, those that echo the political campaign that once revolved around "Asian values" (Glassman 2016). In this way, Asia as method is about more than transcending the East-West binary (Chen 2010, 216).

Chen further proposes a new, decolonizing direction for world historiography. According to him, world history is not a history of the Western world and its interactions with its non-Western others, and should not be written as such. Drawing on Mizoguchi's *China as Method* ([1989] 1996), Chen argues that the world that conceives of China as method, for example, is a different world, a multiple polarity, "in that China is an element of its composition … and Europe is also an element" (Mizoguchi [1989] 1996, 94–5, quoted in Chen 2010, 252). For Mizoguchi, as well as for Chen, the study of a place anywhere on earth "impl[ies] one route toward an understanding of world history" (Chen 2010, 253), and, therefore, "the study of China … transcends China proper" (Mizoguchi [1989] 1996, 93, quoted in Chen 2010, 252).

Despite having shed new light on the need for a de-imperialized and decolonized mode of scholarship, the analytic of Asia as method, as developed by Chen and his cohort at *Inter-Asia Cultural Studies*, is not free of shortcomings. Its framework could be understood as a prime example of what Dirlik (2005, 164) called the "Asianization of Asian studies," a movement among Asian scholars who seek to counter the Eurocentric paradigms dominant in the Asian studies field and to usher in the perspectives of Asians themselves about Asian societies and problems in the field. While for Chen (2010), Asia as method is not only about establishing points of reference and connections between different Asian societies but also between ones in "Third World" countries, he does not elaborate on this point in his book. Therefore, the ways in which Asia as method can provide a universal platform for registering a range of historical and contemporary transformative politics beyond Asia remain underexplored. This may be a serious drawback when we reflect on the increasing globalization that penetrates nearly all places of the earth and that has caused similar forms of dispossession and commodification. Furthermore, and in a similar vein, the bounded category of Asia, which is the primary geographical context of Chen's analysis, is far from unproblematic. Harootunian (2012, 18) calls Asia as method a "critical regionalism" where Asia is a political signifier, rather than a cultural one, that can mobilize different dissident and insurgent politics against the assemblage of multiple powers in Asia. Despite its critical signification of Asia, however, Asia as method may still be susceptible to the charge of spatial fetishism, in the sense that Chen does not problematize the notion of Asia itself (its supposed fixity and boundedness), thereby leaving the regionalization of Asia unquestioned (also see Dirlik 2005, 15; Morris-Suzuki 2000).

Moreover, despite all of the promise of Chen's Asia as method as a theoretical construct that helps us rethink the imperialization and colonization of knowledge production, its mode of analysis smacks of methodological nationalism, prioritizing the national scale within Asia as the central unit of analysis and comparison. Baik (2013a) proffers a corrective to this limitation, by rescaling the problematic of Asia as method to the local – that is, to the site or, in his translation, "location" (*hyeonjang*) – in his notion of "core location."

Situating Core Location within the Urban Studies Field

Baik seeks to further push Chen's problematic for the de-imperialization of knowledge production by turning our attention to the contradictions materialized within specific places in East Asia. This effort is not about the revival of the sort of essentialist empiricism that characterized the area studies field in the past, nor is it a reiteration of postcolonialist calls for attention to particularity and a rejection of universality. Baik (2013a, 47) argues that a universal common ground of trans-local resistance can be identified and imagined through core locations. The common ground shared across different core locations promises to be a generative force for a world consciousness, but the core source of this world consciousness stems from the critical reflection of individuals in these core locations on their relations to each other, to their own broader societies, and to people in other places. The sufferings of the people in these core locations are the sufferings of the world, and only by tackling these problems can the world envision and bring about its emancipation (62). Therefore, Baik's concern, while seemingly focused on the local scale, can be understood as an effort to develop a method that helps us move toward a universal ground of solidarity between different people and places. Baik's approach to core location and its political insights also echoes the "standpoint theory" advocated by Marxists (such as Georg Lukács and, of course, Karl Marx himself) and feminists (e.g., Nancy Hartsock, Patricia Hill Collins, Dorothy Smith, and Sandra Harding) who privilege epistemologies, experiences, and praxis of (the most) marginalized and disenfranchized as the telling enunciations of multi-layered power structures and challenges against them (Mohanty 2003, 231–3). Attention to the most marginalized is the most inclusive paradigm for thinking about social justice as well as systemic power (Mohanty 2003, 232).

While Baik's ideas do have their shortcomings (which we briefly address later), we also see that key components of his notion of core location can potentially countervail drawbacks of some versions of postcolonial urban studies works; at the same time, it can still be in line with the project of decolonizing analytic categories and Eurocentric historicism, the key contribution of postcolonial scholarship. As Eom suggests in her contribution to this volume, urban studies has not witnessed much theorization from the standpoint of East Asian cities (for important exceptions to this trend, see Park, Hill, and Saito 2012; Lees, Shin, and López-Morales 2016; Shin, Lees, and López-Morales 2016). Therefore, the notions of Asia as method and core location can provide a method for urbanists who are interested in urbanization in the so-called Global East (Waley 2013), a term coined to challenge the invisibility of East Asian societies within the dominant geographical nomenclatures of Global North and Global South. As a matter of fact, many chapters in this volume try to thread the problematic of core location and Asia as method with a range of theoretical and political questions raised by scholars in the urban studies field over the past few decades, including those in geography and anthropology.

Urbanists who are inspired by postcolonial problematics have contended that the framework of political economy, which has long been dominant in the field, implicitly takes Western cities as the "origin" or the "model" that can explain cities in non-Western societies, and that these approaches often assume an eventual convergence of different cities across the globe – that is, neoliberal cities (Roy 2011). Aihwa Ong (2007) argues that neoliberalism, for example, is a "mobile technology" and is an exception in Asian cities; that is, it is one of *many* forces that shape urban experiences in these cities and, therefore, does not capture the totality of urban processes in these cities, contrary to political economist accounts that often imply a convergence. Other postcolonial urban scholars have also taken issue with the global city paradigm on account of its implicit economism, *a priori* analytical categorization, and supposed Eurocentrism (Robinson 2002; Shatkin 2007). In particular, Robinson (2002) proposes that the analytical expanse of urban research should be extended from global cities to "ordinary cities," which have been rendered "off the map" by Eurocentric urban studies paradigms.

While we agree with the questions that these postcolonial urbanists raise about the universalist frame of political economic urban theories, we contend that a universal common ground of resistant struggles against unjust capitalist exploitation, dispossession, and expropriation that have erupted across different locales in the world still needs to be identified, explained, and highlighted. Postcolonial urban studies have not paid sufficient attention to these issues and the possibility of a universal resistant front against systemic injustices. The episteme of Baik's core location – which starts its analytics from the marginalized places and people that have been oppressed by a range of structural violence, and their resistant actions against complex relations of power – therefore provides an alternative and critical method to the postcolonial urban paradigms that have focused mostly on the discursive challenges to academic Eurocentrism. We emphasize the significance of a pluralistic world view as suggested by postcolonialists, but we also think that multiplying references as a tool to contest Western hegemony may risk falling into the epistemological pitfall of liberal pluralistic thinking, and that a preoccupation with multiplying and pluralizing references can potentially neutralize or bypass historical violence and structural hierarchies. At this point, we want to bring attention to the triad of critical self-reflection, praxis, and communicative connection that Baik posits as the sources by which resistant forces challenge and eventually transform formidable material structures that constantly generate disparities, dispossession, and uneven development.

Therefore, core locations are not only important in revealing the contradictions, disparities, and unevenness that people in the periphery suffer from, but also provide an alternative epistemology for forming a common ground among people and intellectuals across different places who take global transformative politics seriously. While Baik does not explicitly engage with East Asian core locations' relationships to subalterns in the West, our take is that his ideas can still provide a tool for thinking, one through which we can ascertain a common ground that can be formed between subalterns both in the West and non-West. The ideas of core location and Asia as method can help highlight the importance and necessity of inter-referencing between activists and activist scholars based in different places, as a philosophical foundation for scholars who are interested in the question of resistance and praxis. Therefore, rather than viewing this volume as furnishing yet another version of postcolonialism, our aim is to discuss how different places and territories are not sealed and mutually exclusive, and how they are converging on a universal horizon. This universality does not refer simply to trans-local replicability but also to ideas and praxis reverberating across divergent historical-geographical contexts that have emerged in opposition to multiple forms of systemic injustice.

One more issue that we want to raise pertains to how to investigate a location, a site, and a place in an increasingly transnationalizing and globalizing world. We take seriously Palat's (1999) call for a new way of approaching area studies. According to Palat, the decolonization of knowledge production behoves us to question the act of "unproblematically transposing trans-historical categories and historical trajectories" of Western social formations to non-Western ones. But he also contends that we locate and explain the "dense narratives of local processes within larger global forces of transformation" (116). In other words, local processes of change should be theorized in a relational way (vis-à-vis "a wider relational matrix"), whether these are "long term processes of capitalist expansion" or associated broader geopolitical configurations that local processes are integrated within, correspond to, constitute, and transform (116). This also connects to Gillian Hart's (2006) call for a critical rethinking of area studies. Stressing "relational understandings of the production of space and scale," she argues that scholars should heed "the divergent but increasingly interconnected trajectories of socio-spatial change that are actively constitutive of processes of 'globalization'" (981). Drawing on Lefebvre ([1974] 1991), she reminds us that spaces and places are not pre-existing entities, but are *socially produced*, and places should be understood as "nodal points of connection in wider networks of socially produced space" (Hart 2006, 994). What we, scholars who study specific areas, need to illuminate, she urges, is "power-laden processes of constitution, connection, and dis-connection, along with slippages, openings, and contradictions, and possibilities for alliance within and across different spatial scales" (982).

The broader forces and processes that Palat and Hart each discuss are the universalizing processes of (neoliberal) capitalism and its structural power that have synchronized different places with different histories to geopolicial trajectories. Grasping the dialectical dynamics of the local and the global within the capitalist system is not a strength of Chen's or Baik's work, and it is not central in the overall Asia as method school's problematics. These scholars' optic is mostly limited to the realm of ideas and practices of modernity in East Asia and imperialist and militarist violence, including ones related to Cold War and post–Cold War political regimes, but not the ones associated with capitalism and its attendant class, racial, ethnic, and gender oppressions. These limitations certainly circumscribe the analytical and political purview of the concept of core location and Asia as method. The contributors to this volume recognize such limitations and seek to fill this lacuna. They seek to understand in a relational way each location that they examine.

Core Locations in Korea

This volume comprises seven studies regarding different core locations in South Korea. The contributors to this volume are in various ways engaged in "building ideas" (Sun and Yoon 2013) in relation to the problematics addressed in Asia as method and core location, reflecting them in their own research sites. Each study illustrates how a core location is shaped and produced by particular geopolitical and geo-economic histories at neighbourhood, urban, regional, national, and global scales. In particular, the different chapters examine how the multiple layers of geopolitical and geo-economic power that have characterized East Asia are embodied in the terrains of struggles within these core locations. These layers and power dynamics include the legacy of past Japanese colonialism as well as Japan's ongoing economic ascendency in the region; Cold War legacies that are still shaping geopolitical dynamics in the region (e.g., the tension between the People's Republic of China and Taiwan, and the conflicts between North Korea and South Korea; Chinese empires old (before Japanese colonialism) and new (China's soaring economic power in the post-Mao era)); the transpacific ruling class and the US military-industrial complex; the expansion of capitalist regimes in the region, the latest rendition of which is an increasing neoliberalization of countries and increasing circulation of capital and people in the region; and corresponding regimes of racial, gender, sexual, and other oppressions.

Chapters in this volume show how these different geopolitical and geo-economic histories and presents are entangled with each other to effect complex constellations of power and injustices in different core locations in Korea. Furthermore, they also seek to show how these complex constellations of different processes are interconnected to broader global processes – that is, we seek to show how different core locations should be understood as the nodal points of "multiple historical/geographical determinations, connections, and articulations" (Hart 2006, 984). While we take seriously the question of situated knowledge, our vision does not privilege the local scale and difference.

The core location in each chapter is either a physical site of research or a conceptual space, and each contributor offers her own interpretations about the political and methodological significance of that notion. Each chapter also extends the parameters of the notion of core location, by intervening in each scholar's primary knowledge field, whether within home discipline (e.g., anthropology, architecture, geography, urban studies) and/or through thematic problematics in the research site (e.g., ruins, uneven development, foreign aid, solidarity, welfare, fields).

In chapter 1, "The Idea of Chinatown: Rethinking Cities from the Periphery," Sujin Eom examines South Korea's Chinatown in Incheon as a core location, a space rendered peripheral in Cold War East Asia. Historically, the Chinese community in Korea has been disenfranchised by Korea's ethnocentric national citizenship regime. The long history of discrimination toward this population has continuously forced ethnic Chinese to leave Korea, and often the Chinatowns that they had inhabited become derelict spaces. However, with the rise of China's economic power and the establishment of an integrated East Asian economic space, especially from the 1990s, ethnic Chinese and Chinatowns have surfaced as centres of cultural imagination and economic enterprise in Korea. While revisiting feminist postcolonial scholarship's emphasis on unevenness and its discussion of "ruins," Eom argues that both postcolonial studies and Asia as method scholarship need to pay more attention to the growing influence of the People's Republic of China in Asia following the termination of the Cold War in the late 1980s. Eom demonstrates that ethnic Chinese in Korea, who were peripheralized during the Cold War period, are again marginalized in the contemporary new search for Chinatown as an urban economic engine.

In chapter 2, "Seeing the Development of Jeju Global Education City from the Margins," Youjeong Oh examines Jeju as core location through the lens of Jeju Global Education City (JGEC). JGEC is an education-based urban development project initiated by the central government. It houses high-profile international schools and luxurious residential and commercial facilities in an English-speaking environment. Oh asserts that the development process and outcomes of JGEC both represent and reconstitute Jeju's double marginality through the intensified hegemony of English and the dispossession of Jeju's marginalized residents from the land. Engaging with the scholarly literature on "uneven development" and "neo-developmentalism," especially that developed among geographers, Oh examines the long history of dispossession that Jeju has suffered, and how that deprivation manifests itself in complicated ways in the development of JGEC. In particular, she looks into the defilement of Jeju's ecosystem and the commercialization of space during a development process that prioritized developers, well-to-do people mainly from the mainland, and foreign capital, at the cost of the people of Jeju. Oh further points out that JGEC is an exemplary case of South Korea's unquestioned assumptions about English and development, but that, at the same time, the many contradictions that JGEC manifests complicate such desires.

Chapter 3, "Against the Construction State: Korean Pro-Greenbelt Activism as Method," also problematizes the issue of urban development by examining the struggles that unfolded in the 1990s over the deregulation of greenbelt lands in Korea. Laam Hae argues that greenbelt deregulation was a conjunctural outcome of the processes of democratization, decentralization, and neoliberalization in the late 1990s, but also shows how the mechanisms of the "construction state" – a historically sedimented institutional ensemble of the developmental state and Cold War and post–Cold War inter-regional geopolitics – were central to this process. In her examination, Hae engages with the notion of "articulation" as a way to rethink the frame of Asia as method. She further discusses the theoretical and political implications of the notion of core location, which in her case are various greenbelt sites in Korea that were the focal points of struggles waged between the construction-oriented state and environmental activists. She interrogates how examining these contested sites as core locations may help us rethink the postcolonial question. Furthermore, she argues that the particular struggles over the greenbelt that she examines can provide a window through which to view the topography of broader trans-local resistance.

In chapter 4, "Transnational Marriage Migration as Spatio-Temporal Fix in Pohang's Post-Industrial Urban Development through Saemaul," Hyeseon Jeong explores how Asia as method and Marxist theories can mutually expand each other through the case study of Pohang by employing Baik Yeong-seo's twofold-peripheral perspective for an analysis of the transnational intersection of patriarchy and developmentalism. Pohang's housing aid project in Vietnam for the natal families of women who are marriage migrants discloses the fear in Pohang that the economic difficulties of marriage migrants' families might interfere with the city's stability and development. It also shows how international development aid is implicated in amplifying the marginalities of marriage-migrant women while also trying to challenge them. Jeong argues that the notion of spatio-temporal fix (Harvey 2003, 2006), a Marxist concept that highlights the stopgap way in which capital invests in built environments, can be applied to a variety of scale-jumping programs that attempt to provisionally remedy the socio-economic consequences of uneven development, such as South Korea's state-sponsored transnational marriage migration and Pohang's housing aid project in Vietnam for the natal families of women who are marriage migrants. In so doing, this chapter challenges the boundary of a region that is predicated on the idea of inter-referencing in core location and Asia as method, by presenting the unevenness between East Asia and Southeast Asia that is not considered in the extant Asia as method literature.

In chapter 5, "'Locations of Reflexivity': South Korean Community Activism and Its Affective Promise for 'Solidarity,'" Mun Young Cho examines the efforts of grass-roots activists in Korea who have been involved in anti-poverty community development programs in other parts of Asia. This development has been organized by the Korean Action for Overseas Community Organization (a pseudonym) in Seoul. Based on ethnographic research involving veteran activists and younger trainees of overseas development, the chapter interrogates the ways in which the globalization of South Korean community activism seeks to forge international solidarity. Cho highlights the processes in which Korea's veteran activists reflect on their current positionalities vis-à-vis those of overseas anti-poverty activists. For example, the chapter elaborates the activists' reflections on their own role as "double agents" – that is, as front-line activists in the global anti-poverty solidarity movement and, at the same time, project managers of the Korean government's support to aid-receiving nation. Here, Cho engages with the discussion of "inter-referencing" as developed by Asia as method scholars (in particular, Chen and Sun), examining the stories told by these activists, especially the contradictions that these activists recognize in their interactions with communities in the receiving countries.

Continuing the theme of community activism, chapter 6, "The Education Welfare Project in Pine Tree Hill: A Core Location to Assess Distributional and Transitional Forms of Justice," explores neighbourhood activism in Pine Tree Hill (a pseudonym) as a core location of tension between state-led social development and self-(re)generated development of "the social." Jesook Song demonstrates that activism in this particular community arose on account of its being neglected by the post–Korean War developmental regime. Ironically, the community received extraordinary attention after the Asian financial crisis by neoliberal governments that highlighted social development and welfare in order to alleviate the class polarization that resulted from the uneven national growth of previous decades. The Pine Tree Hill community offers significant insights into people's sovereignty in their negotiations with the seemingly benign state and problematizes the ways in which the welfare state and social development promise to deal with the historical and structural unevenness produced by capitalism. In addition to its intervention in the problematic of welfare, this chapter also critically engages in debates surrounding anthropological theories of singularity and universality in conjunction with ideas about Asia as method.

In dialogue with the previous two chapters, which share anthropological interests, in chapter 7, "Situating the Space of Labour: Activism, Work, and Urban Regeneration," Seo Young Park interrogates the plural interpretations of the meaning of "fields" (*hyeonjang* – sites, scene, or locations) of garment labour by different actors, such as grassroots activists, garment workers, policymakers, and ethnographers. By focusing on Changsin-dong as a core location, a neighbourhood near Dongdaemun Market consisting of garment factories and garment workers' residences, this chapter analyses the layers and shifting frontlines of marginality of this neighbourhood. Dongdaemun Market was the hub of the state-led, export-oriented economy in Korea during the 1960s and 1970s, and accordingly became a hotbed of heated labour union activism at that time. But in the post-industrialization period in the 1990s, garment factories became scattered around the city and downsized into small-scale factories in Changsin-dong. In the new century, it has become an emerging site for the city government's new paradigm of urban renewal and rebranding. This change has transformed the relationship between the labourers and their work, and the word "field" has surfaced with different, and often conflicting, meanings and interests among labourers, activists, and policymakers. In examining these processes, Park highlights different temporalities and spatialities enlivened and embedded in this changing labour geography.

Core locations are both field sites and channels through which each contributor engages in a range of problematics, reflecting on the questions raised by the concept of Asia as method. Each study, while it does not explicitly engage with universality as such, reveals clues about the universal state of life and struggles over it, through deep, grounded research. The spirit of this project is about decolonization through self-reflection (*seongchal*), praxis (*silcheon*), solidarity (*yeondae*), communicative connection (*sotong*), and a shared interest in fighting uneven development in Korea and beyond. We hope these explorations mark the beginning of exciting and fruitful dialogues with other critical area studies and transnational scholars.

REFERENCES

Anievas, Alexander, and Kerem Nişancıoğlu. 2017. "Limits of the Universal: The Promises and Pitfalls of Postcolonial Theory and Its Critique." *Historical Materialism* 25 (3): 36–75.

Baik, Young-seo. 2012. "Sinjayujuuisidae Hangmunui Somyeonggwa Sahoeinmunhak: Ssun-geowaui Daedam" [The Intellectual Call in Neoliberal Era and Social Humanities: Dialogue with Sun Ge]. *Dongbangjihak* [Eastern Knowledge] 159: 421–71.

Baik, Young-seo. 2013a. *Haeksimhyeonjang-eseo Dong-asiareul Dasi Mutda: Gongsaengsahoereul Wihan Silcheon-gwaje* [Rethinking East Asia in Core Locations: A Task for a Co-Habitation]. Paju, KR: Changbi.

Baik, Young-seo. 2013b. "An Interconnected East Asia and the Korean Peninsula as a Problematic: 20 Years of Discourse and Solidarity Movements." *Concepts and Contexts in East Asia* 2: 133–66.

Chakrabarty, Dipesh. 2000. *Provincializing Europe: Postcolonial Thought and Historical Difference.* Princeton, NJ: Princeton University Press.

Chatterjee, Partha. 2004. *Politics of the Governed: Reflections on Popular Politics in Most of the World.* New York: Columbia University Press.

Chen, Kuan-Hsing. 2010. *Asia as Method: Toward Deimperialization.* Durham, NC: Duke University Press.

Chibber, Vivek. 2013. *Postcolonial Theory and the Spectre of Capital.* London: Verso.

Dirlik, Arif. 1994. "The Postcolonial Aura: Third World Criticism in the Age of Global Capitalism." *Critical Inquiry* 20 (2): 328–56.

Dirlik, Arif. 2005. "Asia Pacific Studies in an Age of Global Modernity." *Inter-Asia Cultural Studies* 6 (2): 158–70.

Glassman, Jim. 2016. "Emerging Asias: Transnational Forces, Developmental States, and 'Asian Values'." *Professional Geographer* 68 (2): 322–9.

Harootunian, Harry. 2012. "'Memories of Underdevelopment' after Area Studies." *positions* 20 (1): 7–35.

Hart, Gillian. 2006. "Denaturalizing Dispossession: Critical Ethnography in the Age of Resurgent Imperialism." *Antipode* 38 (5): 977–1004.

Harvey, David. 1989. *The Condition of Postmodernity: An Enquiry into the Origins of Cultural Change.* Cambridge, MA: Blackwell.

Harvey, David. 2003. *The New Imperialism.* New York: Oxford University Press.

Harvey, David. 2006. *The Limits to Capital.* London: Verso.

Jameson, Fredric. 1991. *Postmodernism, or, The Cultural Logic of Late Capitalism.* Durham, NC: Duke University Press.

Lees, Loretta, Hyun Bang Shin, and Ernesto López-Morales. 2016. *Planetary Gentrification.* Cambridge, UK: Polity Press.

Lefebvre, Henri. (1974) 1991. *The Production of Space*. Cambridge, MA: Blackwell.
Mizoguchi, Yūzō. (1989) 1996. 日本人视野中的中国学 [China as Method]. Translated into Chinese by Li Suping, Gong Ying, and Xu Tao. Beijing: Chia Renmin University Press
Mohanty, Chandra Talpade. 2003. "'Under Western Eyes' Revisited: Feminist Solidarity through Anticapitalist Struggles." In *Feminism without Borders: Decolonizing Theory, Practicing Solidarity*, edited by Chandra Talpade Mohanty, 221–51. Durham, NC: Duke University Press.
Morris-Suzuki, Tessa. 2000. "Anti-Area Studies." *Communal/Plural: Journal of Transnational and Cross-Cultural Studies* 8 (1): 9–23.
Ong, Aihwa. 2007. "Neoliberalism as a Mobile Technology." *Transactions of the Institute of British Geographers* 32 (1): 3–8.
Paik, Nak-chung. 2013. "South Korean Democracy and Korea's Division System." *Inter-Asia Cultural Studies* 14 (1): 156–69.
Palat, Ravi Arvind. 1999. "Fragmented Visions: Excavating the Future of Area Studies in a Post-American World." In *After the Disciplines: The Emergence of Cultural Studies*, edited by Michael Peters, 87–126. Westport, CT: Bergin and Garvey.
Park, Bae-Gyoon, Richard Child Hill, and Asato Saito. 2012. *Locating Neoliberalism in East Asia Neoliberalizing Spaces in Developmental States*. Hoboken, NJ: Wiley-Blackwell.
Robinson, Jennifer. 2002. "Global and World Cities: A View from Off the Map." *International Journal of Urban and Regional Research* 26 (3): 531–54.
Roy, Ananya. 2011. "Conclusion: Postcolonial Urbanism: Speed, Hysteria, Mass Dreams." In *Worlding Cities: Asian Experiments and the Art of Being Global*, edited by Aihwa Ong and Ananya Roy, 307–35. Malden, MA: Wiley-Blackwell.
Shatkin, Gavin. 2007. "Global Cities of the South: Emerging Perspectives on Growth and Inequality." *Cities* 24 (1): 1–15.
Shin, Hyun Bang, Loretta Lees, and Ernesto López-Morales. 2016. "Introduction: Locating Gentrification in the Global East." *Urban Studies* 53 (3): 455–70.
Sun, Ge. 2003. *Asiaraneun Sayugonggan* [Asia as Thinking Grounds]. Translated by Jun-pil Ryu. Paju, KR: Changbi.
Sun, Ge. 2007. *Dake-uchi Yosimiraneun Mureum: Dong-asiaui Sasang-eun Ganeunghan-ga?* [The Question That Is Takeuchi Yoshimi: Is an East Asian Ideology Feasible?]. Translated by Yeo-il Yoon. Seoul: Grinbi.
Sun, Ge. 2013. "Discourse of East Asia and Narratives of Human History." *International Critical Thought* 3 (2): 215–23.
Sun Ge, and Yeo-il Yoon. 2013. *Sasang-eul Itda: Munhwawa Yeoksaui Gangeugeul Neomeoseon Daehwa* [Building Ideas/Thoughts: A Conversation

Connected through the Gap between History and Culture]. Paju, KR: Dolbegae Publishers.
Waley, Paul. 2013. "Pencilling Tokyo into the Map of Neoliberal Urbanism." *Cities* 32: 43–50.
Yoon, Yeo-il. 2014. *Sasang-ui Beonyeok: Ssun Geoui "Dake-uchi Yoshimiraneun Mureum" Ilkgiwa Sseugi* [A Translation of Thought: Sun Ge's Reading and Writing of "The Question of Takeuchi Yoshimi"]. Seoul: Hyeonamsa.
Yoshimi, Takeuchi. (1960) 2005. *What Is Modernity? Writings of Takeuchi Yoshimi*. Edited and translated by Richard F. Calichman. New York: Columbia University Press.

1 The Idea of Chinatown: Rethinking Cities from the Periphery

SUJIN EOM

One day, he told me to "learn" China. Isn't it funny? It was a Korean who told a Chinese to go to China, to learn China. He continued to say, "If you get to know China well, it will get you money at the end no matter what you will end up with."

– Pan, interview with the author, 5 November 2014

On 18 December 2014, a local newspaper in the Korean city of Incheon ran a story that recounted how the city was building a Chinese history museum – with virtually no input from the Chinese community. The article drew attention to the fact that the Chinese residents of Incheon, in this and other matters, were denied respect and legitimation.[1] A few days after the story ran, I sat together with the chair of a local Chinese association. It was not a formal interview, but I was asking him a couple of questions about an old shophouse in Incheon's Chinese neighbourhood where his family had once lived – a shophouse that, by that time, had been razed.[2] We looked at reddish-brown pictures and maps to

---

"The idea of Chinatown" is originally the title of an article written by Kay Anderson (1987) in which she challenges the conventional understanding of Chinatown that relies on a discrete "Chineseness" and instead argues that "Chinatown" is a white European idea that informs the making of racial categories in British settler colonialism. While acknowledging the significance of her discussion in the study of race, place, and power, I focus on how the idea of Chinatown takes on an entirely different meaning for Chinese immigrants when the idea itself is in circulation.

1 "Hwagyodeul ppajin hwagyo yeoksagwan" [History Museum without Chinese Residents], *Kyeong-in ilbo* [Kyeong-in Daily News], 18 December 2014.

2 A "shophouse" refers to a housing type combined with a commercial shopfront and is commonly associated with overseas Chinese. Shophouses will be discussed in greater detail in the sections that follow.

figure out how those images corresponded with his memories. As the conversation came to an end, Zhu, a physically fit man in his early forties, pressed me, as if he had been waiting for the right moment. "Because I'm the chair, they ask me to voice our opinions about the museum. But can you please tell me, as you have been to many Chinatowns in Japan and elsewhere, how to do it and what it even means to make our voice heard?" (Zhu, interview with the author, Incheon, 22 December 2014).[3] He was pointing to the recent polemic surrounding the lack of Chinese involvement in the planning process of the history museum. How would Koreans normally react under such circumstances, he asked me, and how would people in other Chinatowns respond in similar situations? I tried to answer to the best of my knowledge. After the meeting, however, his questions lingered in my head, questions posed by someone in a community whose voice had never been heard beyond its confines.

This chapter is an ethnography of Incheon's Chinese neighbourhood in transition. It was in 2006 that I first visited the neighbourhood to undertake preliminary research. From 2009 onward, I began my research in earnest by interviewing residents, sitting in on several town meetings, and attending community events. When I returned to the site in August 2014, some of the residents still remembered me and helped reconnect the researcher with the community. During the intervening time, the neighbourhood had undergone significant transformations. When I first visited in 2006, a handful of new Chinese restaurants had begun to fill what was a predominantly residential district; but at that time, there was still plenty of rubble, and ivy creept up the walls of empty buildings. Behind the main street stood a decrepit Chinese restaurant, which had fallen into disrepair quite a while ago, after the owner's family left the neighbourhood. The old brick building, whose decorative facade hinted at its glory days, was standing in neglect, with big chunks of paint flaking off the walls. Few people could be spotted anywhere. What awaited me eight years later was a landscape of striking contrast. Written in simplified Chinese, a slew of placards welcomed tourists from mainland China. A large number of tourists were flocking to the district, even on weekdays. The once-abandoned Chinese restaurant building had been converted into a city-owned museum, repainted and refurbished.

Behind the changing landscape of the Chinatown lay the growing influence of the People's Republic of China (PRC) following the

3 All the names of my interviewees are pseudonyms.

termination of the Cold War in the late 1980s. After long decades of severance, the movement of people, capital, and ideas between once-estranged countries increased in an unprecedented way. Scholars have noted the historical significance of this period, whose epochal transition upon the dissolution of the Cold War brought about a political, economic, and cultural restructuring of the globe. Economic routes were reopened, diplomatic relations were normalized, and "East Asia" was rediscovered as an object of intellectual and cultural production (Sun 2012; Baik 2013). It was in cities, and port cities in particular, that this transitional moment was felt most acutely, as reopened borders resulted in a new influx of investment capital and tourism revenue.

This chapter traces the cultural repercussions in Incheon's Chinese neighbourhood after once-disconnected economies were reconnected. When assessing these impacts, I draw on Ann Stoler's thinking about ruins. Ruins, according to Stoler, do not simply refer to monumental relics or buildings that have fallen into decay, but rather indicate "the material and social afterlife of structures, sensibilities, and things" (2008, 194). The material environment of the Chinese neighbourhood can therefore serve as a reminder of the violent process and indelible traces of the Cold War in East Asia. Arguably, the ruination brought to the space of diasporic Chinese speaks to what Baik Young-seo (2013) terms a "twofold periphery" (*ijungjeok jubyeonui sigak*): not only have regions such as East Asia occupied a marginal sphere in world history, but these regions also have their own places and people cast as "peripheral" to the formation of national identities. As Baik himself sees a possibility of revisiting history from the vantage point of sites that have been made marginal to nation-states, this chapter takes this peripheral space as a point of departure to reflect on the materiality of the Cold War as it is embedded in East Asian cities.

However, Baik's notion of a twofold periphery should be expanded here so as to articulate the geopolitical complexity of East Asia in the postcolonial world. As I will show in the sections that follow, the peripheral space of Incheon's Chinese neighbourhood can be understood as a consequence of its displacement and dispossession by a postcolonial state that sought to define its national identity within the Cold War structure, an entangled regime of violence to which it was likewise subjected. In the wake of the rise of the PRC in global capitalism over the past few decades, the Chinese neighbourhood garnered increased public attention; yet it led only to another form of ruination in the lives of the people who continue to live with ruins. I will show that the disregard and then sudden appreciation of the ruined landscapes of the Chinese neighbourhood reflect the complex structures of sentiment that are engrained in Korea, where the idea of "Chinatown" as a real and imagined space reveals conflicting sensibilities in the post–Cold War years.

Neither Colonizer, Nor Colonized: The Chinese in the Postcolony

Whether it be from Walter Benjamin's ruminations on the fragility of capitalist culture (Buck- Morss 1991) or Theodor Adorno's musings about ethics ([1970] 1997), "ruin" has long been a famous metaphor for the violent and fragile nature of human civilization. Even after people who once inhabited the city are long gone, the buildings that housed them tend to remain, discharging a different sense of time distinct from the span of a human life. Abandoned buildings may remind observers of memories of the past; yet sometimes the state of abandonment itself also alludes to things, places, and persons displaced to the margins of official history. The past lives on in material forms such as dirt, debris, lichen, patina, and rust, through which the present may be revisited.

Ann Stoler (2008) attends to ruin as "a virulent verb" by highlighting the issue of "mind." "Ruination" is an important term in the analysis of a process that has a corrosive effect on the minds and lives of people who continue their lives in ruins. As opposed to the material environment that is the outcome of abrupt change, ruin can also be engendered by a slow and long-term process that may produce no spectacular images but instead have an attritional impact on human minds. Not only do tangible things perish and become ruined over time, but also the people living with and in ruins are engulfed by material remains in the aftermath of violence, a violence that, paradoxically, leads people to bring another ruination to their built environment by demolishing what is left of it. It is through this co-constitutive process of ruination that people "participate in the making of ruins" (Navaro-Yashin 2012, 152).

In this chapter, I approach Incheon's Chinese neighbourhood as a ruined landscape that occupies a symbolic void in the colonial history of Korea. The term "postcolonial" has long triggered contentious scholarly debates, particularly as to whether the word has lost its critical edge by being a "politically vacuous term" (Choi 1993, 78). Ella Shohat points out that the proliferation of the term also tends to ignore "the politics of the location" (Shohat 1992, 99) – pointing to the homogenizing tendency of postcolonial discourses that do not take into consideration different geopolitical contexts – whereas Anne McClintock contends that discussions of the postcolonial cannot do justice to the global uneven development of postcolonialism (McClintock 1992, 87). If "postcolonial" should instead be employed in a way that enables a scholarly engagement with the geopolitics of ex-colonies, the discursive absence of Japan's former colonies – Taiwan, Korea, Manchuria, Okinawa, and the South Pacific islands – in postcolonial studies deserves attention. As Jini Kim Watson suggests, these ex-colonies of Japan have been ignored in postcolonial studies while being "*over*-privileged by modernisation studies" (2007, 172).

Kuan-Hsing Chen (2010) points out that this Cold War structure of knowledge production "intercepted, interrupted, and invaded" (121) the project of decolonization in East Asia, thereby rendering it incomplete. However, I would argue that, even the attention paid to Japan's problematic role in the post–Second World War years alone does not provide a satisfactory answer to the complexity of the affective topography that is deeply engrained in East Asia, or "local structures of sentiment" (Chen 2010, xiv) that are peculiar to the region. In order to comprehend the "postcolony" as an entanglement of multiple temporalities (Mbembe 2001), one needs to understand the particular historical development of East Asia, where a multiplicity of imperial formations, old and new, have left distinctive marks on human minds as well as material environments. The centuries-long Sinocentric system was replaced by European and Japanese colonial orders in the first half of the twentieth century, which in turn became enmeshed in the Cold War regime by the mid-twentieth century. Meanwhile, modern empires in East Asia have made, unmade, and remade physical and emotional boundaries among people within such a short period of time, which came to produce an intricate mesh of sentiments and meanings in the region.

Building on Ann Stoler's discussion of "ruins" as an alternative vocabulary for the engagement with the tangible effects of imperial formations (2008, 2013), I contend that an interrogation of Chineseness in the postcolony provides a methodological standpoint. In many of the postcolonial nation-states with reconstructed polities in Asia, overseas Chinese became subject to scrutiny and suspicion, due in large part to the in-between economic and political roles they had performed during the colonial era. Given the complex nature of the roles played by Chinese subjects in the advancement of capitalist development in the colonies, either as collaborators of European and Japanese imperialism or as labourers who competed with natives, the Chinese in newly independent Asian countries were often treated as if they represented the residue of colonialism. In Southeast Asian countries, ranging from Malaysia (Loo 2013) to Indonesia (Kusno 2000), the category of "Chinese" thus produced anxiety and uncertainty in the midst of national identity formation after decolonization. As much as imperial projects were inscribed on spaces, postcolonial contradictions also manifested themselves through the material environment (King 2003). The category of "Chinese" often led to physical attacks on Chinese property, thereby revealing the violent process of postcolonial state-formation, which imagined and reconfigured the national citizenry in the form of urban space.

Employing ruins as a vantage point for unearthing the hidden layers of the postcolony, this chapter shows that Chinese Settlements on the Korean peninsula offer a privileged site for examining how different regimes of governance have left traces in the material environment. I differentiate "Chinese Settlements" (*cheongguk jogye*) from the more generic term "Chinatowns" in order to clarify the historical specificity of the former in the Korean context and avoid the uncritical use of the latter that is commonly found in everyday language. Contrary to the popular understanding of "Chinatowns," derived from the experiences of English-speaking countries and circulated across contexts through mobile media such as films and texts, "Chinese Settlements" on the Korean peninsula were the product of legal institutions peculiar to the changing international order in late nineteenth-century East Asia. My use of the term "Chinese Settlements" in this chapter, therefore, emphasizes this particular historical development, whereas "Chinatowns" refer more broadly to spaces imagined and created from exogenous influences. It is almost impossible to relegate these spaces to the confines of particular historical periods such as the mid-twentieth century, especially when we consider the enduring legacies of colonialism in Cold War culture (Kwon 2010). What Lisa Yoneyama (2016, 206) calls "conjunctive cultural critique" of geohistorical violence also points to "the not-so-obvious linkages and connections" among different temporalities. My intention is not to reduce the Cold War and its ongoing effects on mind and matter to that particular period, but rather to fathom the slow violence and protracted process of colonialism in East Asia, which has brought a gradual and attritional ruination to peoples, things, and places.

Chinatown: An Urban Palimpsest of Imperial Formations

Since its establishment in the 1880s, Incheon's Chinese Settlement has existed as a site where different forms of legality were considered normal. The treaty port system, first implemented in Shanghai after the Treaty of Nanking in 1842, introduced the idea of extraterritorial jurisdiction to East Asian ports, from the Japanese port of Yokohama to the Korean port of Incheon, a system that allowed foreign nationals to reside in specially designated "Foreign Settlements" and that granted the residents "near complete immunity" from local laws (Cassel 2012). Under the military protection of the Qing government, Chinese merchants and traders enjoyed economic and political privileges in Korea. The treaty port system was finally abolished after the Japanese annexation of the Korean peninsula, yet the exceptionality attached to Incheon's Chinese Settlement remained intact, and the Chinese continued to lease the land on favourable terms. After the collapse of the Japanese Empire following its unconditional surrender in the Asia-Pacific War, the US military government implemented several policies in an attempt to sever South Korea's trade with Japan and to increase its economic ties with China, including Hong Kong and Macau. Moreover, the military government dealt with ethnic Chinese in South Korea as Allied nationals and consequently provided them with favourable conditions for accumulating capital, particularly in trade (Wang 2005; Lee and Yang 2004). In light of the dominating economic performance of Chinese traders, a Korean-language newspaper lamented that "the night of foreigners" was still in progress at the port of Incheon.[4] This take on things pointed undoubtedly to the port's history – as the treaty port selectively opened to foreign commerce in the nineteenth century as well as the major colonial port under Japanese rule, both of which reminded the Korean public of their loss of sovereignty over the territory.

Along with the Cold War division, nationalistic sentiments in South Korea rendered Chinese residents "foreigners," casting doubt also on their political and economic loyalties. The Alien Landownership Act of 1961 was enacted to prohibit foreigners from acquiring land. The ultimate effect of this legislation was to make Chinese land rights conditional and temporally limited, a far more tenuous position than had applied in the past. In many cases, landownership was recognized only under a condition that basically allowed the state to seize the land for its own use: foreigners' land rights always came with a proviso that, when municipal authorities required the space for the sake of "city planning," the owner must "follow the directions provided." This precarious legal status of landownership rights contributed significantly to the formation of Chinese self-identity in postcolonial Korea and their

4 "Tonggwan ui gihyeonsang: manchu-idong e ggokggok" [A Strange Phenomenon at Customs Clearance], *Kyunghyang Shinmun* [Kyughyang Newspaper], 4 November 1948.

dis-identification with their place of residence. Starting in the 1960s, a range of urban renewal projects (Eom 2018), from road construction to residential developments, further took advantage of Chinese property-owners. A number of Chinese-owned restaurants, shops, and graveyards were demolished, with very little compensation. These actions left the Chinese no choice but to seek economic opportunities elsewhere, such as in Japan, Taiwan, and the United States. In 1972, the number of ethnic Chinese in South Korea was 32,989; by 1982, the figure had fallen to 28,717; and by 1992, it had decreased further to 22,563 (Pak 1986, 210; Lee and Yang 2004, 91).

The out-migration of Chinese residents was closely bound up with how the postcolonial state imagined its national citizenry. In addition to South Korea's Nationality Law, enacted in 1948 on the premise of the principle of *jus sanguinis* and patrilineage, popular nationalism reinforced the gendered national identity by *feminizing* its non-nationals and rendering their presence in the country temporary. (For a discussion of related issues, see Jeong's chapter on marriage migration in this volume.) This process was often manifested in an expectation of their eventual "return" to where they were imagined to have come from. A newspaper article from October 1958 described an international trip of thirty-five Chinese residents to Taiwan to celebrate the Double Ten Day, the national holiday of Taiwan commemorating the Xinhai Revolution.[5] The article portrayed these Chinese travellers as though they were visiting *chinjeong* (a Korean term that indicates a married woman's parents' home),[6] thus likening the trip to an excursion of a married woman visiting her parents to celebrate the birthdays of family members.[7] Chinese residents were gendered as female, while Korea as a nation was implicitly anthropomorphized as male.

This public imaginary displacing the Chinese residents from the national citizenry continued in another form, effacing Korea as the homeland of Chinese residents. A newspaper article in 1962 featured the visit of a Chinese woman named Li Shiu-ying to Incheon, her

5 Starting in the early 1950s, the Republic of China began to actively embrace overseas Chinese due to the regime's competition with the Communist government in mainland China. This was often addressed in the form of supporting trips to Taiwan by overseas Chinese so they could develop a pro-Taiwan national identity.

6 "Chinjeong ganeun hwagyodeul" [Chinese [Women] Residents Visiting Their Parents' Home], *Kyunghyang Shinmun* [Kyunghyang Newspaper], 8 October 1958.

7 This gendered national identity that was inscribed was not just a metaphor: the Nationality Law had not recognized the naturalization of foreigners born to non-Korean fathers.

"second home" as the article described.[8] Li was Miss Taiwan 1961 and the first runner-up in that year's Miss World pageant, and it was her first visit to her family after the beauty contest was held in November 1961 in London. Although Li was born in Incheon and raised by parents who came from Shandong Province in mainland China, Taiwan was unequivocally portrayed as her "homeland."[9] No one seemed to bother to question the complex identity of the overseas Chinese in South Korea, shaped by domestic and international Cold War politics. Further, it was hardly conceivable to live well as a Chinese in an anti-communist state, a country that had cut off the physical and emotional connections that the Chinese community had maintained with mainland China.

Incheon's Chinese neighbourhood gradually fell into decay. Many of the Chinese residents moved out and on to other countries. But, as Lin recounted, only those who had the means of moving out actually left. For those left behind, the bleak landscape of the Chinatown was a blunt reminder of Cold War ruins and ruination – immobility, isolation, and exclusion. The old shophouses disappeared one by one, and rubble and ivy came to inhabit those that remained. In 1984, one of the largest restaurants in Incheon's Chinatown, Gonghwachun, was finally shut down after long years of management difficulties. Many of the restaurant owner's descendants had already migrated to Taiwan for a new life by the time it was closed. People of power, wealth, and intellect emigrated, leaving behind only those with no means of upward mobility (Lin, interview with the author, 24 November 2014). While the causes of and motivations for re-migration varied, the residents who remained were overtaken with the sentiment of "being left behind."

Lin was born in Incheon's Chinatown in the early 1960s and has lived there ever since. She could not speak much Korean until after her graduation from high school, simply because she did not have to learn Korean as long as she lived in the Chinese district. Lin's inability to speak Korean in her childhood, notwithstanding her upbringing in a Korean city, suggests the extent of social distance that existed between Chinese residents and the rest of Korean society. After quitting her first job at a travel agency targeting tourists from Taiwan and Hong Kong,

8 "Oneul je i gohyang-e miseu chaina iyang" [Miss China Li Visits Her Second Home Today], *Dong-a Ilbo* [Dong-a Daily News], 25 January 1962.

9 According to an estimate by a Korean newspaper, as of August 1962, the number of Chinese residents from Shandong was 21,924, accounting for over 90 per cent of the total Chinese population in South Korea (24,029). "Seoul jogye (6): Hwagyoui hyeonhwang" [Seoul's Concession (6): The Present of Overseas Chinese], *Kyunghyang Shinmun* [Kyunghyang Newspaper], 3 August 1962.

Lin worked as a translator at a maritime transportation trade union, where she translated the correspondence with a Taiwanese port city. It was not until after her experience of working for a Korean organization that she learned the Korean language and made Korean friends. It was her first official encounter with Korean society. The formal employment by the Korean firm also earned her a national medical insurance card for the first time in her life.

As a "foreign resident," Lin had been unable to use the medical care system of Korea, and she described her new access to that system as "unthinkable" for most Chinese residents. The Chinese were subjected to a different mode of governance, characterized by differential access and rights to life itself. "What if the children fell ill?" I asked Lin. "Moms took care of the kids. What else could they do? Fortunately, we were healthy most of the time. We had no right to be sick," she responded. The postcolonial condition of the Korean state along with the Cold War division also rendered it almost impossible to live freely as Chinese. Police surveillance continued within the neighbourhood. "Do not stand out" was an unspoken rule, a mantra for the Chinese. Kids learned from an early age that they ought not to chatter in the Chinese language while on the bus or in public space. Directly or indirectly, they might have heard from their families or village elders about numerous incidents in which "Chineseness" – associated with the use of their mother tongue and a "dubious" loyalty to the economic regime of the postcolonial state – would provoke attacks and land seizures.

Shophouses: The Forgotten Architectural Heritage of Chinese Migrants

"It was like an island," said Chen, reflecting on her childhood in an area near Incheon's Chinatown (Chen, interview with the author, 24 November 2014). The social isolation of the Chinese community during the Cold War decades can be glimpsed in a survey taken in the early 1980s, where over half of the respondents (59 per cent) stated that they interacted only with Chinese people in their daily life (Pak 1986). Going through the onslaught of the postcolonial state's nation-building project, Incheon's Chinatown had gradually faded away from the public memory. Oh Jung-hee's autobiographical novella *Chinese Street* (1979), whose English translatation was published in 2012 under the title *Chinatown*, is among the few works that convey the bleak and desolate landscape of Incheon's Chinese neighbourhood of that time: "Those Western-style houses were strange, their steeply slanting roofs and pinched ridgelines looking out of place with their bulk. Perched on a hill that stood alone like a distant island amid the swarm of people on their way to the wharf, they radiated an air of cool contempt. Facing out to the sea, their orifices closed tight like shells, they seemed somehow heroic even in the shabbiness. How old were they? What history did they contain?" (Oh [1979] 2012, 35–7). What Oh describes as "strange" and "shabby" houses were actually a building type known as Chinese shophouses. In the early years after the opening of the port in the late nineteenth century, a number of Chinese migrants from Guangdong in southern China came to Incheon and formed what is now Incheon's Chinatown. The Chinese migrants were merchants and traders, but they were also carpenters, masons, and plasterers who came to the new port of Incheon as construction labourers for Western residences and trading houses. It was they who brought the shophouses of treaty-port – or British colonial – style into the Korean port city. A shophouse is a vernacular architectural form of diasporic Chinese communities commonly found in Southeast Asian port cities (Widodo 2004). While its specific layout and architectural style vary by region and the class of its occupants, it typically refers to a building type with a shop on the ground floor opening toward the street, with residential accommodations upstairs (Kusno 2014). A small courtyard inside the building serves as an open space through which air and light are brought into the compound, which is occupied either by multiple households or individuals. Based on its origins in Southeast Asia's Chinese quarters, this architectural form has been transferred elsewhere, including Incheon.

Not only has the architectural history of Chinese shophouses received little scholarly recognition in Korea, but also their presence has remained almost unknown to the public. Chinese migrants helped introduce red brick to the peninsula through the port of Incheon, which then became widespread in the construction of modern edifices. However, the architectural form of shophouses did not travel beyond the Chinese district. Although shophouses managed to survive the Korean War, their numbers began to dwindle by the 1970s, as Chinese families left the neighbourhood. The disappearance of shophouses was a ramification of the symbolic violence exerted upon the Chinese during the Cold War decades, which "denied [them] a place in the ethnonation" (Lee 2004, 80) and forced them out of what was construed as the "national" space. It was the product of state violence exerted upon Chinese residents, a form of violence legitimated in the name of the nation. Even those buildings left in the neighbourhood were, as Oh Jung-hee portrays aptly in her novella, "closed tight like shells." This left the outsider with access only to the "appearance" or facade of the shophouse building. The inside of the shophouses, and, by extension, the lives of Chinese residents, had long remained obscure. And there was probably no one in the 1970s who could have imagined that these shabby houses in a peripheral location of a Korean city would merit academic research. Similarly, hardly anyone in those years could have imagined that the bleak and desolate neighbourhood would be re-evaluated as part of Korea's heritage.

It was, perhaps ironically, almost impossible even for the Chinese residents to appreciate the shophouse as part of their architectural heritage. The houses, which already contained multiple units, came to be further divided to make more rooms to accommodate low-income families. Even courtyards, which provided open space for ventilation and sunlight to mitigate inconveniences in such a congested environment, were covered and converted into rooms. Born and raised in Incheon's Chinatown, Lin remembers her upbringing in a shophouse-type building in the 1970s, where multiple households lived together sharing a small courtyard. Zhu, having grown up in Chinatown until the 1980s, still has pictures of a shophouse that his family had shared with five or six other households. Both of these shophouses were demolished in the 1980s and 1990s.

As a response to the once-disconnected and now-reconnected market after the rapprochement between mainland China and South Korea in 1992, these Chinese neighbourhoods came to receive public attention. The city government of Incheon accordingly began to invest in the development of its Chinatown, a neighbourhood with worn-out houses and few people on the streets. What should a Chinatown look like? No one seemed to have clear answers. Study tours seemed to provide the city officials with easy solutions that they might learn from precedents elsewhere. In order to formulate Chinatown images, the city officials began to visit Chinatowns abroad and collaborate with Chinese city governments in Shandong Province.

The shophouses began to emerge as an architectural heritage of Incheon's Chinatown. Buildings that had long stood in neglect, the shophouses now seemed to give the area a unique character. In 2000, the city government held a design competition for the Chinatown and its vicinity, where a large number of buildings from the early twentieth century remained. Through the competition, the city wanted to exhibit "models of modern architecture by national characteristics" in order to create a unique street atmosphere and "a street museum" that would display different architectural styles (Jin 2006, 180). The preservation value of a building was determined on the basis of whether 1) it was considered significant at the city and national level; 2) it related to major historic events of the city; and 3) it represented a distinctive architectural style. Buildings not in harmony with surrounding urban structures or "related to the modern history of the city but weak in characteristics" were deemed "insignificant." Buildings that appeared to be at odds with these preservation norms were regarded as "poor architecture" (*bullyang geonchuk*), which should be improved through the application of proper design guidelines so the neighbourhood as a whole could convey a coherent, distinctive appearance (Jin 2006, 188–9).

After the design competition, a Korean architectural firm was selected to take up the mission of formulating design guidelines to refurbish the Chinatown, creating a commercial street named "China Mall." The selected design team derived what they termed "distinctive facade elements" (*oegwan teukseonghwa yoso*) from the front sides of shophouses built in the early twentieth century, which included slanted pitched roofs, "*cheongpung*" (Qing-style) awnings, protruded pilasters, balconies with balustrades, arched shutters, and lattice windows. These elements were meant to be applied uniformly to the facades of other buildings to form a coherent "Qing-style" appearance, the period to which they attributed the history of the Chinese neighbourhood (Jin 2006, 187–91).

Decisions regarding what is to be preserved as heritage are fundamentally questions of how to remember and narrate the past. The making of heritage is not simply an act of imposing values on what is perceived as historical and at risk of erasure; rather, as Jane M. Jacobs has succinctly pointed out, it is a process whereby certain artefacts and places are "incorporated into sanctioned views of the national heritage" (Jacobs 1996, 35). In the city government's plan to preserve historic buildings within the Chinatown, the migrant history of the diasporic Chinese, which has never comfortably fallen under the category of single nation-states, was given meaning *only when* it had to do with the modern history of the city of Incheon and the Korean state. What the design guidelines also reveal is that the transnational nature of the buildings was reduced to stylistic dimensions that would express "national characteristics," but it is questionable which characteristics count as such. In order to represent the "Qing-style" atmosphere, the city government encouraged the use of blue pavement materials within the Chinatown, based on the Chinese character of "qing," which is used to refer to the Chinese dynasty (清) and the colour blue (青) alike. Such decisions raise numerous questions. What does the "Qing-style" atmosphere entail? Is it to Qing China that most Chinese residents today have emotional attachments? And why should it be Qing China that the contemporary Chinatown should resemble?

Aside from these questions, which must not have been discussed among the city officials or designers in the first place, one may dismiss this genre of historical approximation as an instantiation of blithe insensitivity to different cultures. Moreover, the architectural refurbishing of the Chinatown did not proceed in methodical manner but instead by fits and starts. It was often subject to the whims of city officials or the victim of poor communications with the Chinese cities. The arch at the entrance to the Chinatown, which was first erected in 2000 through a donation from the Chinese city of Weihai, has undergone a process of construction and destruction three times. The design guidelines for the facades of shophouses and street furniture have changed a number of times in a capricious manner, thereby creating visual inconsistencies in the neighbourhood.

The long-forgotten architectural form of shophouses was perceived as heritage to be preserved, yet it was only the appearance of buildings that seemed to matter. This could be understood as the city government's facile attempt to commercialize architectural heritage, evidence of lack of sufficient academic research on its architectural history, the social distance between Chinese residents and the Korean government, or a combination of these. But I would argue that this is, rather, a reflection of the sheer level of confusion and ambiguity felt by people since the end of the Cold War, when the idea of Chinatown was mobilized under the influence of a sudden influx of information, ideas, capital, and imageries. Not only were the city officials and designers confused about what a Chinatown should look like, but, more importantly, the Chinese residents themselves had to "learn" about Chinatowns after long years of disconnection from their homeland and isolation from Korean society. When I asked if he had any pictures from the time his family lived in a shophouse, a Chinese restaurateur in his fifties said, "I wish I had one now! It is unfortunate that we did not recognize back then it would become a valuable heritage of ours today" (Liu, interview with the author, 24 August 2014). The visual display of Chineseness, however defined, now appeared as a lucrative business opportunity.

Learning Chinatowns in the Midst of Ruins

Incheon was not alone in municipal attempts to refurbish or even newly construct Chinatowns in Korea (Eom 2017). The reconnected economies that emerged upon the rapprochement of the formerly antagonistic countries, as well as the neoliberal economic restructuring after the financial crisis, fostered different imaginations of Chineseness. Overseas Chinese (*huaqiao* in Mandarin; *hwagyo* in Korean) in South Korea have come into focus as transnational "entrepreneurs" who could help build a bridge (*qiao*) to the Chinese market and capital. While construction of the new Chinatowns is a spatial strategy to attract Chinese capitalists from overseas, ethnic Chinese residents in South Korea are given the role of intermediaries, interpreters, or agents who are expected to eventually "forge a strategic alliance with other ethnic Chinese as well as with – and by extension – China."[10] This punning use of the Chinese word *qiao* ironically reveals the changing role of overseas Chinese in global capitalism. Once regarded as "residual Chinese," overseas Chinese subjects have become "triumphant moderns" (Ong and Nonini 1997), not only for the Chinese state itself but also for other states such as South Korea vying for their investment capital. At the peak of this transition came the repeal of the Alien Landownership Law in 1998, which had long restricted the landownership of foreigners, namely ethnic Chinese. Further, in 2002, the Permanent Residency System was introduced for long-term residents of Chinese descent in Korea, who had long been excluded from the national polity. The subjectivity of overseas Chinese was reconfigured, as they came to be seen as transnational agents capable of connecting local sites to Chinese global capital. Following this line of thought, the idea of Chinatown has emerged as a platform to bridge various markets at a distance.

In the midst of the nationwide development boom of Chinatowns, Chinese residents in Incheon experienced mixed feelings. When Wang heard the word "Chinatown" in the early 1990s used to describe the neighbourhood in which he had spent most of his childhood, he felt strange (Wang, interview with the author, 13 May 2009). His reaction was partly because the residents had their own name for the neighbourhood – *xijie*, a Chinese word that literally means the west street – but, even more, because the English word "Chinatown" as a tourist destination sounded odd to them, when they had already seen many of their better-off Chinese neighbours leave the community. Was there anything in the neighbourhood that deserved to be visited and seen? What was left behind was a handful of old shophouses that had been converted into either cheap lodging houses or laundries. Many of the buildings that used to exhibit the prosperity of the Chinatown – trading offices, restaurants, inns – had already fell into ruin. It was not merely the built environment that appeared at odds with the idea of Chinatown as a tourist destination. Who could truly revitalize the neighbourhood after all the people with the means and power to do so had already gone?

While the idea of Chinatown sounded like an oddity at first, the nationwide Chinatown fever came to have an impact on Chinese residents in Incheon, for whom China as well as "Chinatown" became an object

10 "Ethnic Chinese Move to Build Chinatown in Seoul," *Korea Times*, 12 February 2000.

Figure 1.1 Buildings in Incheon's Chinatown had long stood in neglect. Photographs by author, 2006–9.

of learning. This changed orientation had to do with a new economic space that opened up for the Chinese. Before normalization in 1992, they engaged mainly in small-scale trade between South Korea and Taiwan. But as the port of Incheon began to reconnect to the Chinese port cities of Shandong Province, business ties between the two countries increased. Courted as "bridges" to connect the economies, the Chinese began to see more economic opportunity in their home villages in the Chinese province. Pan, an Incheon-born Chinese businessman, was among those who developed economic ties to Shandong. He had not learned Korean until his twenties, in response to his personal resentment toward Korean society. Although he picked up some elementary Korean language from comic books in his childhood, he was not able to speak much Korean until after he was employed as a translator for a Korean legal firm.

These changing circumstances dovetailed with the increasing disillusionment of the second-generation Chinese who had regarded Taiwan as their "homeland." After graduating from high school in Incheon, quite a few moved to Taiwan to receive a college education and ultimately find employment. However, it was not the homeland they had long hoped for. They felt out of place even more severely than they had in Korea. This feeling of being a stranger in what they had long dreamed of as their homeland was enormously painful, as Pan recounts:

> They always said to me, "You Koreans." I had vaguely imagined that they would embrace me like a mother if I went to Taiwan. But it was not true. I spent ten years in anguish after returning from Taiwan. Later on, I came to realize that "national boundaries" are something that is artificially drawn and "where I am *now*" is my country and home. It was in Incheon that I felt "at home." I don't know why. After all, Incheon is where I was born and everything looks familiar. I hated Koreans, but at least I could understand their behavioural patterns. I had no clue in Taiwan, let alone China. (Pan, interview with the author, Incheon, 5 November 2014)

After his return from Taiwan, Pan engaged in various occupations, from running a Chinese restaurant to intermediary trade with Taiwan and Hong Kong. One day in the early 1990s, a Korean acquaintance advised him to "'learn' China." "Isn't it funny? It was a Korean who told a Chinese to go to China, to learn China," Pan recounted. "He went on to say, 'If you get to know China well, it will get you money at the end no matter what you will end up with.' He was proved right." In 1995, Pan moved to Weihai in Shandong province. Like other Chinese residents in South Korea, he was able to speak the Shandong dialect and thus easily tapped into the Chinese market and bureaucracy. He did whatever he could in order to win the favour of Communist Party officials: he cleaned their houses and emptied the garbage. After five years of hard work, Pan acquired knowledge of the "system" there. He accumulated wealth by running logistics companies in Weihai and came back to Incheon to open restaurants in the Chinatown. He even designed the interior of his restaurant himself, decorating it with furniture, artefacts, and paintings that he had imported from the mainland.

Along with their shifting cultural identities, the nationwide Chinatown fever enabled the Incheon Chinese to rethink their own neighbourhood. It is significant to note that this was taking place in the aftermath of displacement. The idea of Chinatown became equated with a new influx of capital, ideas, and people to the once-abandoned neighbourhood. The residents came to see their Chinatown as an opportunity to bring those who had left back into town. Wang, a Chinatown-born university lecturer I interviewed in 2009, displayed mixed feelings over the changes brought to the neighbourhood. He did not like the city government's attempt to commercialize the area as if it were a theme park, but he still held a faint hope that, with "development," the neighbourhood might be able to bring back those who had left. (On a similar theme, see Oh's chapter in this volume on the development project of English education in Jeju.)

Wang was not alone in his hope of bringing people back. Lin expressed a similar desire, recollecting her classmates from the Chinese school. Many of her friends, who are now in their fifties and sixties, live outside of Korea. She showed me many pictures that she had taken with her schoolmates at reunion gatherings, which were held in various locations ranging from Incheon to Taiwan. One of her friends who had moved to Taiwan for a college education wound up becoming a journalist and settling in New Taipei after marrying a Taiwanese man. Another had moved to the United States and started up a Japanese restaurant in Chicago. "My friends often told me," she said, gazing at the pictures, "they want to come back."

The strong longing to bring back former residents, combined with the anticipation of economic benefits, came in the form of aspirations for "development." Upon hearing of the new atmosphere, a developer visited the neighbourhood to propose a development plan for Incheon's Chinatown. His plan was simple: he wanted to build a multi-storey shopping mall to attract and meet the needs of tourists from mainland China. On the proposed site stood a Chinese church built in 1917 by Chinese missionaries. In order to build the mall, the church had to be demolished. The size of the church was small, but the building represented an early twentieth-century style of religious architecture. There were objections to the demolition, especially among local Korean artists who appreciated the historic value of the religious site: the church had stood in the neighbourhood for over eighty years, making it the oldest Christian church in the city.

Central to the campaign for demolition was an Incheon-born pastor of Chinese descent. A Korean documentary, which aired in 2003, documented the process by which, in order to solicit public support for the demolition, he visited every household in the Chinatown.[11] Before the decision was made to demolish the building, about thirty church-goers gathered to vote on whether the building should be demolished for development or preserved as community heritage. Only one person, a Korean photographer who had taken pictures of Chinatown's ruined landscape, made a strong objection, speaking of the fear of losing the architectural heritage for good. Everyone else voted to demolish the church. They agreed that demolition would be the only way to develop the neighbourhood, thereby bringing prosperity to the Chinese community and

11 "Nihao chainataun" [Hello, Chinatown], *Korean Broadcasting System*, 4 February 2003.

enabling it to grow. More importantly, they hoped that the development of the Chinatown would bring people back to the neighbourhood.

The historic church, whose building retained the memory of Chinatown, was razed in 2002 and replaced by a multi-storey building; however, the financial promise of the new development was not as great as originally anticipated. It did not take long for the community to realize what they had lost. The community denounced the pastor for the deed and he was ousted from the church, the neighbourhood, and even the Chinese Christian parish in Korea. The demolition of the church reflected both the anxiety and aspirations of the Chinese residents. Paradoxically, the creation of what they imagined to be a Chinatown appeared possible only through the act of demolishing what they already had.

This creative destruction unveils the complexity of the Chinese diasporic community, whose sense of uncertainty about the future has complicated their relationship to the built environment. This self-awareness regarding the built environment did not occur in a vacuum, but in the midst of a repetitious cycle of construction and destruction. Playing a significant role in perpetuating this self-initiated process of creative destruction was the unprecedentedly active involvement of the Korean city government, whose indifference to and ignorance of the material consequences of the Chinatown development on Chinese residents continues to cause ruination to the built environment. (For related discussion of Korea as a "construction state" with respect to urban development, see Hae's chapter in this volume.) A Chinese resident denounced the city government's whole plan to develop Chinatown, declaring that they were in fact making the Chinese community "accessories" to Korean society, something they could put on and take off whenever they wanted (Pan, interview with the author, Incheon, 5 November 2014).

After the connections were remade between mainland China and South Korea, the Chinese came to acknowledge the idea of Chinatown as a way to improve their lives. "Chinatown," the place where they have long lived, suddenly became a subject they had to learn about. The Incheon Chinese started thinking about what Chinatown would mean to them and developed remarkably varied ways of making sense of their relationship to the built environment after long years of agonizing over their "place" in the world. This self-awareness has taken place in the midst of changing circumstances that required them to become, all of the sudden, agents who could "bridge" the cultural and economic connections between the two countries. As Zhu recounted, they were always told "not to stand out" during those years of closed-off borders. With the change being so abrupt, they were now left with ambiguous feelings about their shifting positions, which prompted Zhu to ask me what they should to make their "voice" heard and what it would even entail.

Conclusion

Through a close look at Incheon's Chinese neighbourhood in transition, this chapter has discussed a peripheral space in South Korea as a core location through which to examine the materiality of the Cold War as embedded in a local place. The double marginality inherent in its landscape reveals the enduring structure of sentiment that permeates the material environment, a structure of feeling that South Korea has developed vis-à-vis its once-imperialist and communist neighbour during and after the Cold War years. While scholars have noted the historical meaning of the post–Cold War years, the change was particularly significant for diasporic peoples such as Chinese residents in Incheon, who had long suffered from the Cold War division of East Asia. Driven by the nationwide Chinatown development, the idea of Chinatown sparked a sense of self-awareness in the diasporic Chinese regarding their built environment, and they, in turn, began to act upon their material space in the hope of bringing prosperity to the Chinese community that had yet to grow. The city government's facile attempt to commodify the neighbourhood was a reflection of the sheer level of confusion and ambiguity felt by people after once-disconnected cultures were reconnected. More importantly, the Chinese residents themselves also had to "learn" Chinatown after long years of disconnection from their homeland. The confusion often found expression in destructive forms, revealing a paradoxical circumstance that the creation of what they imagined to be a Chinatown appeared possible only through the act of demolishing what already had its own existence.

Incheon's Chinese neighbourhood existed as a space of exception over the course of the twentieth century, where different logics of governance defined its normalcy and in turn created distinct material traits in the built environment. In the midst of contemporary Chinatown development projects across the country and their biopolitical aims to attract Chinese "bodies" devoid of historical specificity, Chinese residents developed their own way of trying to make sense of the change, yet still remaining baffled and confused by it. What Zhu sought to ask me about using the language of "voice" was this form of confusion over the caprice of sovereign rule, a confusion resulting from the gradual and attritional effects of state violence that disproportionately afflicted marginal subjects within the Cold War regime.

The invisibility of Chinese subjects in the official history of South Korea does not only reveal the contradictions inherent in its self-understanding; Incheon's Chinatown also renders visible the state violence responsible for the ruination of people's lives. Neither fully colonizer nor entirely colonized, the Chinese have maintained an almost ghostly presence on the Korean peninsula over the course of the twentieth century, in the shadows of colonialism and the Cold War. I would argue that the ruined landscape of Incheon's Chinatown should not be read as an isolated case confined to a particular locality but rather as an optic through which to understand how multiple temporalities are entangled to produce complex sensibilities within a postcolonial state.

REFERENCES

Adorno, Theodor W. (1970) 1997. *Aesthetic Theory.* Minneapolis: University of Minnesota Press.

Anderson, Kay. 1987. "The Idea of Chinatown: The Power of Place and Institutional Practice in the Making of a Racial Category." *Annals of the Association of American Geographers* 77 (4): 580–98.

Baik, Young-seo. 2013. "An Interconnected East Asia and the Korean Peninsula as a Problematic: 20 Years of Discourse and Solidarity Movements." *Concepts and Contexts in East Asia* 2: 133–66.

Buck-Morss, Susan. 1991. *The Dialectics of Seeing: Walter Benjamin and the Arcades Project.* Cambridge, MA: MIT Press.

Cassel, Pär Kristoffer. 2012. *Grounds of Judgment: Extraterritoriality and Imperial Power in Nineteenth-Century China and Japan.* Oxford: Oxford University Press.

Chen, Kuan-Hsing. 2010. *Asia as Method: Toward Deimperialization.* Durham, NC: Duke University Press.

Choi, Chungmoo. 1993. "The Discourse of Decolonization and Popular Memory: South Korea." *positions: asia critique* 1 (1): 77–102.

Eom, Sujin. 2017. "Traveling Chinatowns: Mobility of Urban Forms and 'Asia' in Circulation." *positions: asia critique* 25 (4): 693–716.

Eom, Sujin. 2018. "Infrastructures of Displacement: The Transpacific Travel of Urban Renewal during the Cold War." *Planning Perspective* (December).

Jacobs, Jane M. 1996. *Edge of Empire: Postcolonialism and the City.* London: Routledge.

Jin, Rin. 2006. "Incheon Gaehanggi Geundaegeonchungmul Bojeon Mit Jubyeon Jiyeok Jeongbi Bang-ane Gwanhan Yeon-gu" [The Preservation of Historic Buildings and Economic Renewal in Jung-Gu, Incheon Metropolitan City]. *Geonchuk gwa Sahoe* [Architecture and Society], 178–93.

King, Anthony. 2003. "Actually Existing Postcolonialisms: Colonial Urbanism and Architecture after the Postcolonial Turn." In *Postcolonial Urbanism: Southeast Asian Cities and Global Processes,* edited by Ryan Bishop, John Phillips and Wei-Wei Yeo. New York: Routledge.

Kusno, Abidin. 2000. *Behind the Postcolonial: Architecture, Urban Space, and Political Cultures in Indonesia.* London: Routledge.

Kusno, Abidin. 2014. *After the New Order: Space, Politics, and Jakarta.* Honolulu: University of Hawai'i Press.

Kwon, Heonik. 2010. *The Other Cold War.* New York: Columbia University Press.

Lee, Jeong-hee and Pil-seung Yang. 2004. *Chainataun Eomneun Nara: Han-guk Hwagyo Gyeongje-ui Eojewa Oneul [A Country without a Chinatown: Yesterday and Today in the Overseas Chinese Economy of Korea*]. Seoul: Samsung Economic Research Institute.

Lee, Jin-kyung. 2004. "National History and Domestic Spaces: Secret Lives of Girls and Women in 1950s South Korea in O Chong-hui's 'The Garden of Childhood and The Chinese Street'." *Journal of Korean Studies* 9 (1): 61–95.

Loo, Yat Ming. 2013. *Architecture and Urban Form in Kuala Lumpur: Race and Chinese Spaces in a Postcolonial City*. Burlington, VT: Ashgate Publishing.

McClintock, Anne. 1992. "The Angel of Progress: Pitfalls of the Term 'Post-Colonialism'." *Social Text* 31/32: 84-98.

Mbembe, Achille. 2001. *On the Postcolony*. Berkeley: University of California Press.

Navaro-Yashin, Yael. 2012. *The Make-Believe Space: Affective Geography in a Postwar Polity*. Durham, NC: Duke University Press.

Oh, Jung-hee. (1979) 2012. *Chinatown*. Translated by Bruce Fulton and Ju-Chan Fulton. Seoul: Asia.

Ong, Aihwa, and Donald Nonini. 1997. *Ungrounded Empires: The Cultural Politics of Modern Chinese Transnationalism*. New York: Routledge.

Pak, Eun-gyeong. 1986. *Han-guk Hwagyoui Jongjokseong* [The Kinship Characteristics of the Chinese Korean]. Seoul: Han-guk Yeon-guwon.

Shohat, Ella. 1992. "Notes on the 'Post-Colonial'." *Social Text* 31/32: 99–113.

Stoler, Ann Laura. 2008. "Imperial Debris: Reflections on Ruins and Ruination." *Cultural Anthropology* 23 (2): 191–219.

Stoler, Ann Laura. 2013. "Introduction: 'The Rot Remains' – From Ruins to Ruination." In *Imperial Debris: On Ruins and Ruination*, edited by Ann Laura Stoler, 1–35. Durham, NC: Duke University Press.

Sun, Ge. 2012. "Reconceptualizing 'East Asia' in the Post-Cold War Era." In *The Trans-Pacific Imagination: Rethinking Boundary, Culture and Society*, edited by Naoki Sakai and Hyon Joo Yoo, 253–77. Hackensack, NJ: World Scientific.

Wang, En-Mei. 2005. "Migunjeonggiui Han-guk Hwagyo Sahoe: Migunjeong, Junghwamin-gukjeongbu, Han-gugin-gwaui Gwan-gyereul Jungsimeuro" [The Chinese Korean Society under the American Military Rule in Korea: The Relationship between the American Military Rule, the Republic of China and Korean]. *Hyeondae Jungguk Yeon-gu* [Journal of Modern China Studies] 7 (1): 87–132.

Watson, Jini Kim. 2007. "Imperial Mimicry, Modernisation Theory and the Contradictions of Postcolonial South Korea." *Postcolonial Studies* 10 (2): 171–90.

Widodo, Johannes. 2004. *The Boat and the City: Chinese Diaspora and the Architecture of Southeast Asian Coastal Cities*. Singapore: Marshall Cavendish Academic.

Yoneyama, Lisa. 2016. *Cold War Ruins: Transpacific Critique of American Justice and Japanese War Crimes*. Durham, NC: Duke University Press.

2 Seeing the Development of Jeju Global Education City from the Margins

YOUJEONG OH

Introduction

Jeju Global Education City (Jeju yeong-eo gyoyuk dosi) is an education-based urban development project that houses an array of internationally recognized schools and colleges, together with residential and commercial facilities, in an English-speaking environment on the island of Jeju. The Roh Moo-Hyun administration (2003–7) initiated the project to reduce the number of students studying abroad and to prevent money that would otherwise have been spent abroad from leaving the country. As of 2006, more than 45,000 South Korean high-, middle-, and even elementary-school students were studying abroad, a figure 1.7 times higher than the 26,600 overseas students in 2001. Over $4.5 billion[1] was spent on study abroad in 2006 alone – an amount that had more than quadrupled from the $1.06 billion in 2001.[2] The Roh administration asserted its strong determination to stop the outflow of people and money[3] and to address various associated social problems, such as the "goose father" phenomenon[4] and "family breakdown." The central and local governments have poured approximately $2 billion into

1 Throughout this chapter, dollar amounts are in US dollars.

2 *Jeju yeongeo-jeonyong-taun joseong gibbon-bangan* [The Basic Plan for the Construction of Jeju English Town], 2007 *Gungmuhoeui bogojaryo* [2007 State Council Reports].

3 The government-sponsored survey estimated that the country would prevent a foreign currency outflow of between $324 and $540 million, if all twelve planned schools are established in Jeju Global Education City and attract a combined total of 9,000 students who would otherwise go abroad to study.

4 The term "goose father" refers to a father who sends his wife and children abroad to study and flies to see them only during winter and summer vacations. The goose father phenomenon has attracted social and media attention because of family breakups and divorces, and even suicides among the lonely fathers.

constructing the 940-acre city, which contains the following: a school zone; a college zone; an English Education Centre; an Education, Culture, and Art Centre; and housing and other commercial facilities. The original plan aims to establish a total of twelve elementary, middle, and high schools, accommodating around 9,000 students.[5] The plan also sets out to attract more than ten colleges and universities, which will share a single large campus, designed to be the biggest in Asia. Since the ground-breaking ceremony in 2009, four schools have opened and begun operating in Jeju Global Education City (JGEC): North London Collegiate School (NLCS) Jeju, Branksome Hall Asia (BHA) Jeju, Saint Johnsbury Academy (SJA) Jeju, and Korea International School Jeju.[6] Once all of the planned schools and colleges have been established, JGEC is anticipated to host 23,000 residents, including students, parents, teachers, retailers, and government employees.

In short, JGEC is a central government–led urban construction project in Jeju with an English education theme. The central government designated Jeju as the project site for the so-called English City development. Jones (1997) calls the process of site designation for state projects "spatial selectivity," implying that "the state has a tendency to privilege certain places through accumulation strategies, state projects, and hegemonic projects" (831). Actually, Korea's entire post-war development process has manifested selectivity: known as the Korean developmental state's "picking the winner" policy, it has favoured capital, men, and the greater Seoul area over labour, women, and local regions (Oh 2018). In order to initiate and expedite economic development on the war-torn land, the developmental state adopted strategies to mobilize national resources and concentrate them in a few designated sectors. For example, the developmental state preferred export-targeted industries and,

5 In 2005, it was estimated that the total number of students who went abroad or applied for foreign-language and international schools in Korea had reached 90,000. The 9,000 students that JGEC plans to accommodate were determined to represent 10 per cent of the potential demand for an English education.

6 The Korea International School Jeju is the only public school, and its operation is affiliated with YBM (a private English educational institute). Its curriculum conforms to, and is accredited by, the Western Association of Schools and Colleges; thus its graduation certificate will be acknowledged both in Korea and the United States. Branksome Hall Asia Jeju is a Korean branch of the private school based in Toronto, Canada. North London Collegiate School Jeju is a Korean campus of the British private school. St. Johnsbury Academy Jeju, which has its home school in Vermont, is the newest addition at JGEC. At the three Western schools, the coursework is identical to that taught at the schools in their home countries. At the schools in JGEC, the whole curriculum is taught in English, except for Korean language and history.

therefore, actively promoted labour-intensive light industries such as textiles, shoes, and wigs in the 1960s and heavy and chemical industries in the 1970s. Given the country's natural resource constraints, spatial selectivity and concentration were also evident in the channelling of investments for infrastructure into a few designated areas to maximize efficiency. Strategic spatial selection favoured the Seoul metropolitan area and the southeast coast, which were relatively less ravaged by the Korean War and retained some industrial facilities (B.G. Park 2003). Within such a geographical privileging process, there has been a firm belief in Korea that, with few exceptions, central state investments in fixed assets bring both material and symbolic benefits to the project sites. The construction of industrial complexes (*gongdan*), Five New Towns (*odae sindosi*), and the Multifunctional Administrative City (now Sejong City) are prime examples of how the state's infrastructure-oriented urban development brought population growth, increased tax bases, and fostered the notion of "development" (*baljeon*), at least within the project areas.

In line with developmental urbanization focused on fixed assets, then, could JGEC be interpreted as yet another state hegemonic project, in this case privileging Jeju? Would the "English City" not only boost the local economy but also render Jeju a global city? To state the conclusion first, JGEC has been the subject of multiple controversies – relating, for example, to environmental devastation, deficit management, and social exclusion – and has provided only negligible economic benefits to the local society. Although the central government initiated and invested in the urban development project, the JGEC project has functioned mainly as an accumulation strategy for certain capitalist and political factions rather than for the benefit of the people in Jeju. Therefore, theories of spatial selectivity cannot adequately explain the irony of this state urban intervention that has resulted in the dispossession of local land and ecological assets. Why, then, did the state's mega-investment in infrastructure construction not economically benefit the island as a whole? It is because Jeju, the project site, is not an empty place in which state intervention brings about the intended outcomes in plain ways – rather, Jeju's situated realities have caused the project to unfold in unexpected ways.

To explain Jeju's particular historical, geopolitical, and economic contexts, I draw on Baik Yeong-seo's concept of "core location" (*haeksim hyeonjang*), which he defines as a doubly marginalized (*ijungjeok jubyeonui sigak*) location in which multi-scalar and multi-dimensional forces intersect with and contradict each other (Baik 2013). Baik designates Okinawa, the divided Korean peninsula, and Taiwan as core locations because, first, they are parts of the East Asian Other according to the West-centred modernization paradigm, and, second, they are peripherally positioned within the internal hierarchical order of East Asia. Under US military hegemony in the Cold War regime in the Asia Pacific, those locations also share the peripheriness into which the installation of US military bases was forced. Although these locations manifest historical contradictions as a result of being under the influence of the Chinese, Japanese, and American empires, they are not destined to remain marginal, due to their critical position in international politics. Although Baik does not narrow divided Korea into subnational areas, following his logic I consider Jeju a core location because multiple layers of marginality have shaped the island's history. First, Jeju is a part of "still developing" South Korea vis-à-vis the West. Given the histories of colonialism, and the influence of the American Empire and Euro-American hegemony, Korea has hardly occupied a hegemonic position, despite its significant economic development. In this chapter, I argue that the persistent neocolonialism that pushes Korea onto the margins is represented by the dominant power of English. Second, given its physical existence as an island off the southern tip of the peninsula, Jeju is the remotest periphery within the country. The uneven capitalist development strategy that the Korean developmental state adopted has rendered Jeju's peripheral condition even more conspicuous. Nevertheless, Jeju lies at the forefront of Korea's global project, as evidenced, for example, by the Jeju Free International City development (discussed below). As a strategic military location and one of the "New Seven Wonders of Nature," Jeju is also representing Korea on the international stage. Such intricate realities of Jeju, as a core location, create multiple tensions that cannot be captured by a single analytical framework.

While Baik's concept of core location is pertinent to situate Jeju's multiple historical tensions, Jeju as a core location further extends the conceptualization. This chapter examines how the JGEC development demonstrates Jeju's condition as a core location, mediating its historically constructed and multi-layered marginalities and recent global aspirations. The next four sections are structured around the concept of core location. In the first and third section, I show the ways in which JGEC exhibits the endeavours of both Korea and Jeju to overcome their marginalization, as witnessed by the postcolonial desire to move up the hierarchical global order by way of developmentalism. In the second and fourth section, I reveal how, ironically, the developmental aspirations embodied in JGEC exacerbate the country's and the island's marginal positions. Baik Yeong-seo claims that a core location is a site of both struggle and practice; thus, the importance of a core location lies in the ways in which the dual marginal position drives people's critical contemplation and practices. As detailed below, the assessment of whether the JGEC development is "successful" or not is divided at both the national and local levels, as there have been sustained critical voices against the project despite the seeming social consensus regarding its necessity. Because the complicated tensions surrounding the JGEC project cannot be adequately revealed through the dispossession/resistance framework only, this chapter attempts to critically extend Baik's concept of core location.

On the Hegemony of English in South Korea

As the Roh Moo-Hyun administration asserted, JGEC is designed to counter the phenomenon of early – that is, pre-college – study abroad (*jogi yuhak*). The rise of (early) study abroad is grounded in the promise of modernity, as migrants leave the "small country" of South Korea for the wider world in order to exploit the gap between those two worlds and achieve upward mobility back home (J. Park and Lo 2012; Besnier 2011). Although overseas education can be viewed as an alternative form of human development, one enabling students to escape from Korea's constricting and oppressive education system in a more democratic and creative environment elsewhere (Kang and Abelmann 2011), a more critical reason for choosing to study abroad lies in accumulating the socio-economic capital that English proficiency and a foreign diploma confer. Considerable financial resources are invested in study abroad in order for individuals to acquire the symbolic value that raises their status in Korea. The mechanism underpinning transactions between financial and symbolic capital is the modernist idea of "linear development," which typically views the West as at an advanced stage. Global economic disparities endow English with a form of power that can be converted into socio-economic capital in Korea, a less advanced country.

The gap between developed and developing countries has long existed across the globe and has recently widened as a result of widespread neoliberalization. Despite South Korea having substantially bridged the gap over the past decades through vigorous economic development, the country is a peculiar place where English has enjoyed an inordinate level of hegemonic power throughout its recent history. In *English, Colonialism in My Mind*, Yoon Ji-Kwan (2007) suggests that the power of English in South Korea stems from the country's subordination to the United States. Korea was never a colony of the West, and thus the use of English was never coerced; however, the country's modernization process was deeply entangled within post–Second World War American hegemony. The US military occupation (1945–48) that ruled the southern half of the country immediately following Korea's liberation from Japanese colonial rule formed the basis of US influence. Because the US-puppet government of Syngman Rhee (1948–60) employed pro-American and US-educated elites in its postwar process of nation building, the privileging and dominance of English became institutionalized in South Korea. The hegemony of English was amplified during the 1950s and 1960s, due to the influx of US aid, which deepened the country's dependent condition. The export-oriented economic development of the 1970s and 1980s valued those who were fluent in English, thereby further increasing the power of the language.

Globalization – arriving in South Korea more by state action than by the unfettered expansion of free market capitalism – has, since the mid-1990s, further deepened the dominant power of English. In the early 1990s, the Kim Young-Sam administration (1993–98) proclaimed a national globalization project, *Segyehwa* (globalization), which placed new emphasis on economic liberalization and deregulation. By drawing a strong link between English fluency and the country's globalization, the *Segyehwa* policy affected English education to a great extent. Extra-curricular English education was instituted for fourth, fifth, and sixth graders in elementary school in 1994, and in 1997 English became a mandatory subject for students in the third grade and above. The introduction of "native teachers" (*woneomin seonsaengnim*) – foreign teachers whose native language is English – into classrooms also began in the 1990s. Undergirding such extensive English initiatives was the state's intention to create a bilingual workforce that could serve the needs of an increasingly global market. Keeping up with the government policy, private after-school tutoring and English learning at private institutions became ubiquitous (S. Park and Abelmann 2004). Sponsored by local governments, English Village (Yeong-eo maeul), a theme park/English school hybrid, began to spread throughout the country after 2004. English Village provides a short-term English-immersion experience in an English-only environment, where students learn how to use English in a variety of contexts.[7]

7 English Villages began with the aim of acting as an inexpensive substitute for study abroad, and, as of 2011, a total of twenty-two English Villages were operating across the country. Having run at a deficit for several years, however, some of them have been closed and others have been privatized (and admission fees raised).

At the height of the government initiatives, the Lee Myung-Bak administration (2008–13) attempted to implement a nationwide "English Immersion Program," according to which all classes would be taught in English by 2010. Although this policy failed in the face of strong opposition from societal forces, it once again drove the whole country into an English frenzy. Reflecting requests from the upper middle class, the Lee administration also dramatically increased the number of foreign schools (*oegugin hakgyo*) and international schools (*gukje hakgyo*) in Korea, claiming that English was necessary tool for success in a highly globalized world. President Roh Moo-Hyun also asserted the importance of English as a means to accelerate Korea's global integration, as exemplified in his plans for the "English City": "English is a must in order to catch up with the globalization stream ... Koreans spend over $10.5 billion annually on private English study lectures and programs, but they are still considered weak in English. The government will gradually expand investments in social infrastructure for English education."[8] All presidential claims deem English an indispensable necessity for Korea's global orientation and individuals' success in the globalized economy. Thus, English serves as a tool in the state's project to cultivate human capital that can enhance its national competitiveness in the global economy.

The state's goal of nurturing global talent through English is also associated with individuals' desires to be competitive human capital in the neoliberal economy. As English fluency acts as a critical evaluation criterion for getting into university, landing a job, and receiving promotions, English education has served as an effective measure of upward class mobility in Korean society since the Korean War. In the aftermath of the financial crisis in the late 1990s in particular, the neoliberalization of the country turned the English fever into a site in which individuals prepare for and practice neoliberal personhood, or the "entrepreneurship of the self." In this view, continuous self-improvement is celebrated as an opportunity for maximizing the value of one's human capital (Kim 2012). The neoliberal reforms of the economy since the financial crisis have dismantled state protectionism by opening South Korea up to greater global competition. This economic restructuring has also set up a flexible labour system in which previous seniority-based promotion and lifetime employment for white-collar workers was replaced by merit- and

Also, sceptical views have persisted about their cost-effectiveness and the practical effectiveness of these short-term stays.

8 "Roh Pledges Aggressive State Investments in English Education," *Hankyoreh*, 7 April 2007, http://english.hani.co.kr/arti/english_edition/e_national/201411.html

performance-based job contracting. Since English is one of the central means through which to evaluate one's performance, it becomes a prerequisite for South Koreans' survival in an extremely tough job market (J. Park 2011). English, consequently, acts as an oppressive power that few South Koreans are free to ignore. As "the ability to communicate in the global lingua franca serves as a recognised means of enhancing one's value to globally conscious employers" (Chun and Han 2015, 567), English proficiency simultaneously provides individuals with abundant resources they can mobilize to promote themselves as assets in the struggle to keep up with the global current. The state invites individuals to cultivate more cosmopolitan capital through English-based overseas experiences, while also exploiting them as free or extremely cheap labour in the state-driven works of human resource–oriented foreign aid. In sum, in a context where global economic disparity is translated into a source of social mobility in Korea, English acts as both an oppressive power forcing individuals constantly to practise it and as a resource to brand oneself as global talent. Such a form of Foucauldian governmentality fosters particular neoliberal subjects, all driven to engage in steady self-development to become or remain competitive human capital (Foucault 1990; Jesook Song 2007; Kang and Abelmann 2011; J. Park and Lo 2012).

Reproducing Western Hegemony

JGEC interweaves multiple aspirations, all mediated through English: the government project of nurturing a global citizenry; the social aspirations toward betterment; and the individual project of self-development aimed at cultivating globally competitive human capital. The Korean name of JGEC is Jeju yeong-eo gyoyuk dosi, the direct translation of which is "Jeju English Education City." This signifies that, while JGEC hosts general education institutions, its practical and symbolic foci still remain in English education. Nevertheless, the creation of JGEC differs from previous policies regarding English education. Transcending mere soft programs and curriculum, JGEC is a "physical intervention" with the goal of creating an exclusive English-speaking city. Despite its call for the domestication of English education, however, the "English City" construction is built on the same assumption as early study abroad – that English competency is impossible to acquire unless one lives in an English-speaking society. In the hope that the artificially controlled environment will have the same effect as studying in a foreign land, JGEC promotes an all-English policy, or English as a common language (*yeong-eo sang-yonghwa*). In accordance with this policy, all students use English as the medium of their education and as a communal language;

indoor and outdoor signs are in English; administrative services are offered in English; native English-speakers are invited to settle in the city; and the number of foreign employees will be increased. Other residents, including parents and retailers, are encouraged to use English, although it is not mandatory.

As a critical part in the creation of a "West-like" environment, JGEC has tried to attract Western schools. With the clear intention to bring in Western institutions rather than develop domestic ones, it has hosted three Western schools and is also engaging in (troubled) negotiations with more foreign colleges and universities.[9] At the ground-breaking ceremony for North London Collegiate School Jeju, the vice-minister of the Ministry of Land, Transport and Maritime Affairs remarked, "Setting up leading international schools in Jeju will enable global-level knowledge acquisition and English learning without going abroad."[10] Although JGEC asserts that it provides a domestic supply of West-like (English) education, it offers the physical venue but procures the educational content from Western institutions. As part of the all-out effort to invite Western schools to open an Asian branch in JGEC, the establishment of for-profit education institutions is allowed. Schools in JGEC are not subject to a tuition cap, so their fees are notoriously expensive, much higher than other private schools in Korea (specific numbers are discussed below). In March 2015, moreover, the central government announced plans for a legal revision that would allow foreign schools in JGEC to distribute retained earnings to their home countries.[11] The

9 Actually, Jeju Free International City Development Center (JDC) provided the first two foreign schools, BHA and NLCS, with free land and covered the construction costs for their grandiose campuses, which were designed by world-class architects. Moreover, when there is an operational deficit, Haewul, a service provider established by JDC to support the operation of schools in JGEC, is supposed to make up the deficit. Given concerns about recouping its initial operational deficit, JDC is reluctant to offer construction and operation costs to late-coming schools. Probably for that reason, many of JDC's attempts to attract more Western institutions have failed during the negotiation process.

10 "NLCS jeju haksaengdeureun geullobeol lideoro seongjanghal geot" [Students at NLCS Jeju Will Be Grown as Global Leaders], *Jeju Sori*, 30 September 2011, http://www.jejusori.net/?mod=news&act=articleView&idxno=105045&sc_code=1395808068&page=156&total=3135

11 "Gukjehakgyo donbeori gwonhaneun jeongbu" [The Government Is Promoting the 'Profit Making' of International Schools], *Hankyoreh*, 15 March 2015. http://www.hani.co.kr/arti/society/schooling/681711.html. Since opening in 2010, the UK's North London Collegiate School and Canada's Branksome Hall Asia were still operating at a deficit until 2015, when talks about the legislative revision emerged. The proposal of the legislative revision is intended to address this problem. While

South Korean regulation had banned the transfer beyond the schools' boundaries of profits they generated. Such a regulation was based on the belief that education serves the public interests, despite a certain level of privatization. The recent move to endorse the profit repatriation, however, signifies that the priority is more to attract Western institutions than to support the publicness of education.[12] As it promotes profit repatriation, the inflow of Western institutions, and the lifting of the tuition cap, JGEC appears to be a showcase for the neoliberal marketization of education in Korea. Moreover, JGEC's initial intention to create an exclusive English City detached from the Korean local context pushes Jeju toward becoming a mere container for a collection of foreign schools. Jeju provides its land, and foreign institutions simply drain profits out of the island and the country.

The construction of the exclusive English City, the promotion of an all-English policy, and the hosting of Western institutions reveal the peripheral character of Korea. The national aspiration to speak English well reflects the country's postcolonial desire to bridge the gap between Korea and the West. At the same time, English is the means to bridge the global disparity because it is the globally hegemonic language. Thus, the position of English in postcolonial Korea creates an ironically self-reinforcing cycle through which the hegemonic power of English deepens. Through urban construction, the Korean state has provided an alternative place to receive a "global-level" education. The actual global education service, however, is not self-sustaining, but is guaranteed only through localizing Western institutions. Thus, the very promise to bridge the temporal and spatial distance between core and semi-periphery actually reproduces the global hierarchy in which Korea is still placed in a subordinate position vis-à-vis the West. JGEC reaffirms the hegemonic power of English and the supposed superiority of Western education, even though it has the effect of retaining domestic students who would otherwise head overseas. The physical existence of JGEC reflects and materializes the global hierarchy, informing Koreans (particularly Jeju locals) about the superiority of English, both implicitly and explicitly. It is even more conspicuous to Jeju locals who are struggling to keep their unique local dialect. The desires built

---

the Jeju Office of Education has officially announced its opposition to the proposal, criticizing it as a means to turn JGEC into a market-driven free economic zone, no other counterviews have emerged from the provincial government, the provincial council, or the national assembly.

12 Further ramifications of the profit repatriation include the possibility of other domestic schools to gradually break up the regulations on the profit transfer.

up in the postcolonial period are (re)producing a form of neocolonialism that subordinates Korea in the hierarchical global order. The reproduction of Western hegemony simultaneously discloses the marginality of Korea.

Jeju as Korea's Global Frontier

In addition to providing a "West-like" English education, JGEC as a central government–initiated urban development project ostensibly also aims to benefit the province. During the search for a feasible construction site, Jeju was "designated" as the project site as it contained a huge chunk of land owned by the provincial government, and thus readily available for the state's use. Due to its physical location, JGEC has become interwoven with a bigger project, the "Jeju Free International City" (JFIC) project, which is intended to develop Jeju into an East Asian centre of tourism, education, and health services, modelled after Singapore and Hong Kong. JFIC is defined as "a regional unit wherein regulations will be relaxed and international standards will be applied so as to ensure the international movement of people, commodities and capital as well as convenience in business activities to the maximum extent."[13] JFIC is formulated as a series of physical development projects: Jeju Science Park, Myths and History Theme Park, Healthcare Town, and Jeju Global Education City.[14] These flagship projects are consonant with other cities' tactics in the "art of being global" through the speculative construction of places, such as creating signature buildings, displaying signature designs, and holding hallmark events (Harvey 1998; Kong 2007; Ashworth 2009).[15] Global cities are usually characterized by their integration into the world economy, thus acting as basing points or nodes for the global operations of transnational corporations, financial services, and production (Brenner and

13 *Jejuteukbyeoljachido seolchi mit gukjejayudosi joseong-eul wihan teukbyeolbeob* [Special Act on the Establishment of Jeju Special Self-Governing Province and the Development of Free International City], proclaimed 21 February 2006.

14 Originally, eight mega-projects were designed. Due to the lack of investment and other issues, currently, the number of projects has been reduced to four, and one public project (Jeju Public Rental housing) has been added. Detailed information can be accessed at Jeju Free International City Development Center, at https://english.jdcenter.com/main.cs

15 Although such drives to "be global" face criticism with respect to the normalization of the global city and the exacerbation of socio-spatial polarization, particularly among "developing" cities, "playing the global" has been a desperate initiative for emerging cities and has legitimized "the entrepreneurial turn" in urban governance.

Keil 2006; Olds and Yeung 2004; Sassen 1991). For those aspiring cities that are not equipped with such functional economic qualities, the construction of urban spectacles is often taken as the best strategy for pursuing global city status. The JFIC development projects, therefore, can be interpreted as Jeju's desire to be listed in the global city roster by creating a spectacular built environment that could attract investment, industries, and tourists. I would like to stress that, although the investigation of the JFIC development fits in with the new trend in urban studies that has identified the complex urban process beyond Euro-American-centred theorization (Shin, Lees, and López-Morales 2016), JFIC still manifests the strong desire to be globally recognized.

JGEC is situated within the larger context of JFIC and is designed to act as an "East Asian education hub." Although more than 90 per cent of currently enrolled students are Korean nationals, JGEC aims to eventually recruit foreign students and their parents, making it truly international. JGEC also intends to compete with Singapore, Hong Kong, and Malaysia, which have already asserted their status as regional education hubs (Olds 2007; Cheng, Cheung, and Yeun 2011; Koh and Chong 2014). Joining in this inter-Asian competition to become a regional hub for international education, Jeju has also developed education as one of its "new growth engines" (*sin seongjang dongnyeok*). Actually, the Special Act for the Development of Free International City defines education as one of Jeju's "future industries." Taking JGEC as the first phase of the JFIC project, the JFIC developers have expressed their aspiration to build a new economic base, advancing its "internationalization" (*gukjehwa*) by attracting students from all over the world, and thus eventually achieving global city status. As a part of the larger JFIC project, JGEC carries out the procurement of individuals equipped with foreign language fluency, with the goal of eventually turning Jeju into a borderless territory with English as its lingua franca.[16]

I interpret JFIC's efforts to form a nexus between JGEC as a regional education hub and the global city formation as a combination of the central government's neoliberal experiments and local development desires. First, JFIC is an "exceptional space" in which multiple regulatory exceptions are applied (Ong 2006). As identified in its definition, JFIC is the materialization of neoliberal globalization policies, given

16 The immediate policy discourses around making English a common language in Jeju faced opposition by the local society, and thus have tentatively been discarded. Yet English is imposed as a common language within JGEC, and the same policy will be gradually expanded into the entire territory of Jeju.

its promotion to the "maximum degree of free movement of people, goods, and capital," seeking to be a "no-visa, no-tariff, no-regulation" zone.[17] The Jeju provincial government already offers visa-free entry for foreign travellers for up to thirty days. The province has also promised a radical degree of deregulation by lowering corporate tax, currently at 13–25 per cent in South Korea, to less than 13 per cent, thereby undercutting Shanghai's Pudong (15 per cent), Hong Kong (17 per cent), and Singapore (22 per cent). Invested firms' labour control would be tolerated within JFIC. The visa-waiver policy, tax incentives, and flexible deregulation are designed to attract and facilitate the influx of foreign capital.[18] The Jeju Special Self-Governing Province introduced "real estate incentives" to enable the liquidity of foreign direct investment.[19] More importantly, JFIC has claimed to allow for-profit education and health care. In the health sector, foreign and local for-profit organizations can establish health facilities to treat locals as well as foreigners, and overseas medical licences are recognized, which they currently are not in the rest of Korea. JGEC is situated within this context of the neoliberal marketization of education by hosting for-profit (foreign) educational institutions. Jeju is part of Korea, but simultaneously an exceptional frontier in which the state is experimenting with extreme neoliberal forms of accumulation that could offer an alternative to the accumulation crisis on the mainland. Given that the ways in which such neoliberal attempts would impact the local community are not assured, it would be even harder to interpret the JGEC project as spatial selectivity benefiting certain areas. Jeju is, rather, an exceptional space in which capital movements could be guided without the deliberate consideration of the quality of local lives.

In effect, the central government introduced the Jeju Special Self-Governing Province (JSSGP) to govern the JFIC development projects.

---

17 *Jejuteukbyeoljachido seolchi mit gukjejayudosi joseong-eul wihan teukbyeolbeob* [Special Act on the Establishment of Jeju Special Self-Governing Province and the Development of Free International City].

18 While the Roh Moo-Hyun administration promised the central government's extensive financial support to complete the JFIC development, the subsequent Lee Myung-Bak and Park Geun-Hye administrations have been reluctant to finance the projects. Rather, those two administrations redirected the financing source from public support and invited private investment, especially foreign direct investment.

19 Permanent residency will be granted to foreigners who invest over $500,000; longer residence stays will be allowed for foreigners who purchase a condominium over $200,000; and those who invest more than $5 million can acquire casino business rights. Already more than 2 million m^2 of Jeju land has been sold to mainly Chinese capital, and the foreign investors' buying up of Jeju is accelerating.

Effective 1 July 2006, Jeju was designated as a special self-governing province, whereby authority over everything except issues pertaining to diplomatic relations and defence was devolved to the province. Jeju province carried out the administrative breakthrough by dismantling the previous two-city, two-county system and incorporated them into two cities (Jeju and Seoguipo). Having exclusive rights to appoint the mayors of the two cities, the elected provincial governor has come to have substantial power over affairs within the Jeju territory. Thus, Jeju province appears to have achieved decentralized and self-regulating power, at least in the administrative system. Despite its seeming self-governing status, however, it is noteworthy that the ideas and action plans of JSSGP and JFIC are envisioned and implemented by the central government. Thus, it would seem that the administrative revolution was nothing more than a measure to facilitate a series of development projects for promoting JFIC, simplifying and centralizing the decision-making process by leaving it to the provincial governor.

In addition to its status as an "exceptional space," Jeju, like many other postcolonial cities, has deep-rooted developmental desires, which are especially strong given its history of deprivation and marginalization. In the premodern era, the island was a remote place of exile to which political criminals were sent. Cut off from the mainland, Jeju never was integrated into mainstream politics. The proverb "Send any horse to Jeju Island, and send any human to Seoul" captures the peripheralized status of the island. Further marginalization and oppression occurred in the aftermath of the Jeju Uprising of 3 April 1948, when, during the US military occupation, Jeju rose up to oppose the separate elections in the mainland that eventually led to the virtual division of the country (Cumings 2005). Condemned as the "Red Revolts," the uprising, which led to the massacre of as many as 80,000 residents, aggravated the isolation of Jeju. Such ideological oppression of Jeju residents as part of anti-communist nation-building indicates once again how Jeju was a core location marginalized under US military hegemony in East Asia. Its peripheral status was deepened during the post–Korean War economic development process, which was operated on an uneven strategy. During the economic development of the 1970s and 1980s, Jeju was not only neglected but was always forced to be "special." To exploit its warm weather, Jeju people were pushed to grow "special" produce, such as bananas, pineapples, and tangerines, whose values fluctuated owing to their dependence on external markets. In the name of capitalizing on the "especially" beautiful natural environment, massive tourism development occurred in Jeju in the 1980s. Although the "special" policies affecting the island brought relative wealth to its residents, compared to underdeveloped areas on the mainland, they also intensified the deterritorialization of the islanders. Tourism development, implemented through eminent domain, displaced Jeju residents from their land and made them dependent on visitors, further undermining self-sufficient and self-determining endogenous development. The recent construction of a naval base in Jeju also uprooted many of its residents from their land and livelihoods. Such histories of oppression and suffering have instilled in Jeju strong aspirations for development (*baljeon*).

As Jeju is not an undifferentiated whole, the ideas and practices of betterment are represented in diverse forms among locals and those who care about Jeju. Many Jeju-born artists and activists have tried to rediscover forms of local indigenousness such as local dialects, the practice of female divers (*haenyeo*), and distinct local nature and culture. Recent cultural migrants to the province appreciate the "de-growth" (*talseongjang*) narrative and practices such as slow lifestyles as a form of resistance to the fast-paced, compressed modernization of the rest of the country. Promoting Jeju as one of New Seven Wonders of Nature, even the Jeju provincial government has supported efforts to preserve its precious natural environment. Against such diverse imaginations and voices advocating alternative "betterment," what is intriguing is the way in which the state-driven urban development projects become a dominant mode of materializing developmental desires. The central and local governments' policy guidance, which is immediately implemented through its legal justification, attracted private capital from the island, mainland, and overseas, making Jeju a physical site in which construction-oriented developmental energies are animated. The JFIC development can be understood as a combination of the "special" status allocated to it by the central government and Jeju province's prolonged desire to be economically and politically advanced, particularly within the context of the country's decentralization. It is the country's global frontier in which protective and exclusive regulations are relaxed in favour of foreign capital. At the same time, it is a means for the aspiring Jeju politicians and capitalists to break with the island's history of underdevelopment and achieve global advancement. Behind the JFIC project lies a firm belief that the spectacular transformation of the built environment and the influx of foreign capital and foreign nationals will turn Jeju into a global city, thereby overcoming its peripheral and underdeveloped legacy. Developmental spirit is, however, not only about the subaltern "capacity to aspire" from the situation of marginalization (Appadurai 2004). As aspirations have "the power to move and motivate" (Bunnell and Goh 2012), they are also about energy, creativity, and the driving force to make changes toward desirable goals. JGEC manifests diverse local and national players' developmental ambitions, which are intertwined with imagining, performing, and achieving the sense of being global: toward global talent, global citizenry, global education hub, and global city. Here, *the global* is a critical way to brand the city as a desirable place to visit and in which to reside, educate children, and invest capital.

Reinforcing Jeju's Marginality

Have the JFIC and JGEC projects actually turned Jeju into a "global city"? It might be too early to evaluate the final outcome of these projects. Up to the present, however, JGEC has been plagued by various controversies. First, it spawned controversies over "noble schools." Situated in the context of the JFIC's neoliberal marketization of education that allows for-profit (foreign) educational institutions, schools in JGEC are not subject to a tuition cap, as mentioned above, and their tuition rates are notoriously expensive. For example, the annual tuition fees at North London Collegiate School Jeju are KRW 27.67 million ($23,650)[20] for high-school students. If the dorm expenses are factored in, the figure rises to KRW 42 million ($35,900). While it is argued that such expenses are only half of the cost of studying abroad, ordinary Koreans cannot afford the costs of tuition and extras that, when combined, are almost equivalent to the annual total income of ordinary households. Not surprisingly, then, 37 per cent of Korean students at NLCS are from Gangnam, three posh districts in Seoul, while Jeju residents' enrolment rate remains negligible. Thus, questions arise as to why taxpayers' money should be used, and why Jeju's environment should suffer, simply to build "schools for the rich" who are not Jeju residents. Despite the fact that its ostensible raison d'être is to offer global-level education in a local context, JGEC has become a site in which the established social order is protected and reproduced in Korea. It has turned out to be a class strategy through which the upper class consolidates its privilege (Jae Jung Song 2011), rendering class distinctions and social disparities even more visible to not-so-wealthy Jeju locals. The measures taken to remedy the gap between the West and South Korea have, ironically, made the domestic disparities between the haves and have-nots all the more conspicuous.

Second, the construction of JGEC has raised concerns over environmental devastation. Most of the project site for JGEC belongs to the

20 This amount is based on the exchange rate in March 2017.

gotjawal terrain. *Gotjawal* (a combination of *got* meaning forest and *jawal* meaning rubble or rocky soil in Jeju dialect) refers to a primeval forest uniquely formed on the lava terrain. Jeju's *gotjawal* forest spans four large regions mainly located in *jungsanggan*, or the mid-mountain region, and encompasses 12 per cent of Jeju's land. The *gotjawal* areas were once regarded as worthless, not suitable for farming, and thus remained untouched, even throughout most of the twentieth century. Yet the lack of agricultural value fostered the ecological value of the *gotjawal* forests, which have become home to as many as 600 endangered species of plants, insects, and animals, as well as rich kinds of moss, now called the "lungs of Jeju." More importantly, *gotjawal* forests have the critical function of recharging the groundwater supply (which is the main source of drinking water for Jeju's half million inhabitants) by absorbing, filtering, and diffusing rainwater. Given the *gotjawal's* ecological value and its role in groundwater protection, the development of JGEC on one of the four *gotjawal* terrains was bound to result in disputes. Moreover, colonies of red bark oak are widely found in the project area, enhancing the area's excellent range of vegetation and ecology.

The controversy intensified when the Jeju provincial government eased the ecological regulations tied to the project site. All "green land" (*nokji*) in Korea is categorized to indicate its preservation value according to the level of landscape, vegetation, and underground aquifers: category 1 land should be left untouched, category 2 is subject to strict preservation, category 3 to preservation, category 4 is partially developable, and category 5 is developable. Originally, only 30 per cent of the project site of JGEC was evaluated as developable. In 2007, however, the Jeju provincial government initiated a research project to readjust the ecological categories pertaining to the *gotjawal* terrain. The Korea Research Institute for Human Settlements (KRIHS) conducted the research that became the basis for lowering areas designated as category 3 or 4 land to level 4, making them virtually developable land. The adjustment of ecological categories has expanded the developable area in the project site to a minimum of 42 per cent (1.8 million m^2) and a maximum of 50 per cent (2.1 million m^2). It is no small conflict of interest that KRIHS, which originally proposed the development plan for JGEC, also conducted the ecological feasibility study. The ecological deregulation that virtually paved the way for development became open to public criticism, raising concerns that destroying ecological assets would have irreversible results.

Third, JGEC has remained an enclave in Jeju, yielding far fewer positive impacts for the island than expected. The original plan had anticipated that the 23,000 residents newly added in JGEC would boost the local economy. But JGEC has remained isolated – like an island within the island. When I visited the city, I had a hard time finding the site even with the help of GPS, despite being a Jeju native. JGEC was literally "hidden" deep inside hilly areas. I did not see any active economic interactions between areas inside and outside the city. Most families in JGEC assume the domestic form of "wild goose families," with fathers staying and working on the mainland while mothers live in Jeju and care for their children. Given that their major spending is on schooling, I wonder how actively these goose families boost the local Jeju economy. The lack of spillover effects is also evidenced in interviews with local people. Nearby villagers said that at first they welcomed the project in the hope that the development would raise the value of their property. JGEC has, however, become a nuisance to nearby locals, since wastewater from the city has flowed to adjacent villages, threatening the villagers' livelihoods that depend on fishing. In a larger context, the main developer of JGEC, Jeju Free International City Development Centre, is affiliated with the central government (Ministry of Land, Transport and Maritime Affairs) and rarely shares locals' interests. The construction companies that have built the residential units in JGEC are all based on the mainland. While the urban development project may generate some temporary construction jobs, it does not lead to any continuity of employment. And the spatially selected project of JGEC once again marginalizes, rather than privileges, Jeju through the profit-repatriation program. Granting permission for profit repatriation will further detach the benefits from the local context. In sum, while Jeju's land is being utilized for the city's construction, the benefits from JGEC have been deterritorialized from Jeju.

To synthesise the three controversies, the JGEC development has brought about mostly negative outcomes for the island, including the negligible enrolment rate of Jeju native students; the destruction of the invaluable *gotjawal* terrain; the deficit management of the project; and the lack of any substantial transfer of development benefits to Jeju residents. Yet, this does not necessarily mean that the state's massive investments have simply melted into air. The state's project enabled the accumulation of foreign schools, developers, and upper-class residents at the expense of the dispossession of Jeju's public land and the invaluable ecosystem. Because the beneficiaries and benefits of the state project move away from the project site, the concrete ramifications of the "urban project" turn out to be mainly the deprivation of local land and ecology, rather than the prosperity of the local community. This state-orchestrated neoliberal form of accumulation by dispossession (Harvey

2003; Glassman 2006; Levien 2013) highlights the marginality of Jeju – particularly the grass-roots islanders – once again.

Seeing JGEC from the Margins

I have discussed the ironies of the JGEC development from mainly two perspectives. First, JGEC manifests the national aspirations toward a West-centred modernization mediated through the English language on multiple levels: the state's project to nurture global citizens equipped with English proficiency and individual projects of self-development toward globally performing human capital. The state's intervention to provide a "West-like" educational environment, however, has only reinforced the hegemony of English in the country. Also, because of the unchallenged belief that advanced education can be achieved only through attracting Western institutions, the state's project to bridge the gap between the West and Korea only reaffirms the peripheral status of the country. Second, JGEC is designed to develop Jeju into a global city, demonstrating both the central state's intentions and locals' desire for development. Yet the global city project has spawned multiple controversies, such as environmental devastation, deficit management, and low spillover effects. The dispossession of local assets (land and ecological value) has directly led to the accumulation of foreign capital and the consolidation of the capitalist class, thereby deepening the marginality of Jeju.

JGEC embodies the country's blind faith in both English and development (*gaebal*): the firm belief that English fluency necessarily translates into socio-economic benefits, and the undoubted confidence that physical development brings socio-economic betterment.[21] In this chapter, I have shown how these two assumptions are easily disputed, demonstrating that, ironically, JGEC has disclosed the marginalities of both Korea and Jeju. The situated realities of JGEC as a core location – in which multiple historically and geopolitically conditioned marginalities are condensed – reveal to us why state-invested urban projects further deprive rather than privilege Jeju.

21 The word "develop" can be translated into Korean in two ways: *baljeon*, meaning "bringing advancement," and *gaebal*, meaning "improving the economy" or "the built environment," or both. The conventional understanding of development in the post–Korean War developmentalist period is that *gaebal* (the urban/industrial infrastructure construction for economic development) brings *baljeon* (betterment) (Oh 2018), and JGEC also manifests such a notion.

The JGEC development is still ongoing, so it may be too early to evaluate the ultimate outcomes of the project. Multiple parties, however, have voiced their perspectives. Complimentary voices have come out from the developers themselves. For example, the Jeju Development Centre has released press reports stating that most students in the first class of NLCS Jeju were admitted to world-class universities, including Stanford and Oxford. The real estate developers are circulating the same rhetoric as a means to sell their condominiums in JGEC to upper-class parents. Critical attention should also be paid to the unthinking manner in which students' advancement toward top-tier Western universities counts as "success."

The sources of oppositional voices, on the other hand, are more diverse. The Korean Teachers and Education Workers Union has argued that the English-immersion education in JGEC is shattering the basis of public education; liberal politicians and critical intellectuals have denounced the government's approval of for-profit education and profit repatriation in JGEC; and local activist groups have decried the damage to the *gotjawal* forests.[22] Since the development project did not involve the direct displacement of residents, ordinary islanders' reaction to the JGEC development appears to be rather lukewarm, in the same way that JGEC remains an enclave in the middle of nowhere on the island.[23] Despite the central-local power imbalance that gave the provincial government a marginal role during the JGEC development process, the provincial government is more eager to promote the project due to the developmental project's physical existence, thereby enhancing the province's accumulation potential. Despite its paradoxical role in doubly marginalizing Jeju, JGEC has not entailed any substantial collective resistance or critical practices, as Baik Yeong-seo's core location theory implies. Even the local opposition to the environmental devastation faded away after the construction was completed. Yet, diverse voices have been continuously heard. The tension among the conflicting perspectives sheds new light on JGEC's character as a core location in demonstrating how state-imposed developmentalism is often highly contested.

22 Citizens' groups include Gotjawal People (www.gotjawal.com), the Gotjawal Trust of Jeju (www.jejutrust.net), and Love Gotjawal (part of the National Trust).

23 Public hearings are held to communicate with locals over the development project. However, local people are not fully informed of such hearings, and thus the venues are usually filled only with proactive activists, who are often prevented from attending by police power. Therefore, most public hearings often end up as one-day events, rarely resulting in a true reflection of local voices.

REFERENCES

Appadurai, Arjun. 2004. "The Capacity to Aspire: Culture and the Terms of Recognition." In *Culture and Public Action*, edited by V. Rao and M. Walton, 59–84. Palo Alto, CA: Stanford University Press.

Ashworth, Gregory. 2009. "The Instruments of Place Branding: How Is It Done?" *European Spatial Research and Policy* 16 (1): 9–22.

Baik, Yeong-seo. 2013. *Haeksimhyeonjang-eseo Dong-asiareul Dasi Mutda: Gongsaengsahoereul Wihan Silcheon-gwaje* [Rethinking East Asia in Core Locations: A Task for a Co-Habitation]. Paju, KR: Changbi.

Besnier, Niko. 2011. *On the Edge of the Global: Modern Anxieties in a Pacific Island Nation*. Stanford, CA: Stanford University Press.

Brenner, Neil, and Roger Keil. 2006. *The Global Cities Reader.* New York: Routledge.

Bunnell, Tim, and Daniel Goh. 2012. "Urban Aspirations and Asian Cosmopolitanism." *Geoforum* 43 (1): 1–3.

Cheng, Yin Cheong, Alan Cheung, and Timothy Yeun. 2011. "Development of a Regional Education Hub: The Case of Hong Kong." *International Journal of Educational Management* 25 (5): 474–93.

Chun, Jennifer Jihye, and Ju Hui Judy Han. 2015. "Language Travels and Global Aspirations of Korean Youth." *Positions: Asia Critique* 23 (3): 565–93.

Cumings, Bruce. 2005. *Korea's Place in the Sun: A Modern History*. New York: W.W. Norton.

Foucault, Michel. 1990. *History of Sexuality*. Vol. 1, *An Introduction*. New York: Vintage Books.

Glassman, Jim. 2006. "Primitive Accumulation, Accumulation by Dispossession, Accumulation by 'Extra-Economic' Means." *Progress in Human Geography* 30 (5): 608–25.

Harvey, David. 1998. "From Managerialism to Entrepreneurialism: The Transformation of Urban Governance in Late Capitalism." *Geografiska Annaler. Series B, Human Geography* 71 (1): 3–17.

Harvey, David. 2003. *The New Imperialism*. New York: Oxford University Press.

Jones, Martin R. 1997. "Spatial Selectivity of the State? The Regulationist Enigma and Local Struggles over Economic Governance." *Environment and Planning A* 29 (5): 831–64.

Kang, Jiyeon, and Nancy Abelmann. 2011. "The Demonstration of South Korean Precollege Study Abroad (PSA) in the First Decade of the Millennium." *Journal of Korean Studies* 16: 89–118.

Kim, Eleana. 2012. "Human Capital: Transnational Korean Adoptees and the Neoliberal Logic of Return." *Journal of Korean Studies* 17 (2): 299–328.

Koh, Aaron, and Terence Chong. 2014. "Education in the Global City: The Manufacturing of Education in Singapore." *Discourse: Studies in the Cultural Politics of Education* 35 (5): 625–36.

Kong, Lily. 2007. "Cultural Icons and Urban Development in Asia: Economic Imperative, National Identity, and Global City Status." *Political Geography* 26: 383–404.

Levien, Michael. 2013. "The Politics of Dispossession: Theorizing India's 'Land Wars'." *Politics and Society* 41 (3): 351–94.

Oh, Youjeong. 2018. *Pop City: Korean Popular Culture and the Selling of Place.* Ithaca, NY: Cornell University Press.

Olds, Kris. 2007. "Global Assemblage: Singapore, Foreign Universities, and the Construction of a 'Global Education Hub'." *World Development* 35 (6): 931–1098.

Olds, Kris, and Henry Wai-Chung Yeung. 2004. "Pathways to Global City Formation: A View from the Developmental City-state of Singapore." *Review of International Political Economy* 11 (3): 489–521.

Ong, Aihwa. 2006. *Neoliberalism as Exception: Mutations in Citizenship and Sovereignty.* Durham, NC: Duke University Press.

Park, Bae-Gyoon. 2003. "Territorialized Party Politics and the Politics of Local Economic Development: State-led Industrialization and Political Regionalism in South Korea." *Political Geography* 22: 811–39.

Park, Joseph Sung-Yul. 2011. "The Promise of English: Linguistic Capital and the Neoliberal Worker in the South Korean Job Market." *International Journal of Bilingual Education and Bilingualism* 14 (4): 443–55.

Park, Joseph Sung-Yul, and Adrienne Lo. 2012. "Transnational South Korea as a Site for a Sociolinguistics of Globalization: Markets, Timescales, Neoliberalism." *Journal of Sociolinguistics* 16 (2): 147–64.

Park, So Jin, and Nancy Abelmann. 2004. "Class and Cosmopolitan Striving: Mothers' Management of English Education in South Korea." *Anthropological Quarterly* 77 (4): 645–72.

Sassen, Saskia. 1991. *The Global City.* Princeton, NJ: Princeton University Press.

Shin, Hyun Bang, Loretta Lees, and Ernesto López-Morales. 2016. "Introduction: Locating Gentrification in the Global East." *Urban Studies* 53 (3): 455–70.

Song, Jae Jung. 2011. "English as an Official Language in South Korea: Global English or Social Malady?" *Language Problems and Language Planning* 35 (1): 35–55.

Song, Jesook. 2007. "'Venture Companies,' 'Flexible Labor,' and the 'New Intellectual': The Neoliberal Construction of Underemployed Youth in South Korea." *Journal of Youth Studies* 10 (3): 331–51.

Yoon, Ji-Kwan, ed. 2007. *Yeong-eo, Nae Maeumui Singminjuui* [English, Colonialism in My Mind]. Seoul: Dangdae.

3 Against the Construction State: Korean Pro-greenbelt Activism as Method

LAAM HAE

On the afternoon of 30 July 1999, a protest and counter-protest spectacle unfolded in Seoul Station Plaza, located in Seoul's central business district. Five hundred protesters from all over the country were railing against legislation that the Kim Dae-Jung government had just announced, an act that mandated the revision of greenbelt laws in the direction of deregulation. They demanded the resignation of the head of the Ministry of Construction and Transportation (MOCT), as well as the annulment of the new act. These protesters were social and environmental activists from about 160 activist organizations that constituted the umbrella group People's Action for Saving Greenbelts (PASG, Geurinbelteu Salligi Gungmin Haengdong). This was an unusual scene, one in which the riot police were guarding protesters, who were chanting anti-government slogans, from the counter-protesters, who were physically and verbally assaulting the activists. Anger, even fury, was intense among the counter-protesters, and Seoul Station Plaza was filled with the smell of the broken eggs that they had thrown at the protesters. The counter-protesters comprised mainly residents and landlords in the greenbelts, most of whom were members of the Korea Greenbelts Association (Jeon-guk Gaebaljehan-guyeok Juminhyeophoe), which had persistently demanded the abolition of greenbelts in Korea.[1]

This protest and counter-protest spectacle (and many similar ones over greenbelt deregulation) was a moment that expressed the contradictions of unfolding post–Cold War geopolitics conjoined with

This research was supported by the Social Sciences and Humanities Research Council of Canada through funding from the Major Collaborative Research Initiative entitled "Global Suburbansims: Governance, Land, and Infrastructure in the 21st Century (2010–2017)."

1 This description of the protest is based Bae and Lee (1999).

political democratization and economic neoliberalization in Korea. Such struggles over greenbelts are the focus of this chapter. I examine the politics of greenbelts from the 1970s onwards and, in particular, the protest actions that ensued in the late 1990s as a result of the state's announcement of the deregulation of the country's greenbelt. In examining this issue, I also engage with the question of how we understand a place and a place-based oppositional politics, drawing on the insights of Baik's (2013) idea of "core location."

First, I argue that, while the neoliberal economic turn in Korea in the 1990s triggered greenbelt deregulation, such deregulation should also be analysed as a moment within the longer history of the peculiar capitalist political modernity that has existed in Korea since the 1960s – that is, the construction state (*togeon-gukga*). The "construction state" refers to a historically specific politico-institutional assemblage that has powerfully shaped the urbanization process in Korea, especially that process's pattern of uneven development. The mantra of "Fight the Construction State" that was chanted by pro-greenbelt activists in the 1990s compels us to chart their organizing struggles within the particular spatial history of capitalism in Korea, as engineered by the construction state and reconfigured through neoliberal processes. The construction state machine, and more recently the country's neoliberal power bloc, has long considered the greenbelt as both *terra nullius* and a disposable reservoir for future development.

Following the emphasis in many postcolonial works, including Chen (2010), I want to highlight the importance of an analysis that attends to geographical and historical contexts and contingencies. But I seek to do this without losing sight of the idea advanced by historical materialist scholars (Palat 1999; Hart 2006; Glassman 2016) that locally particular historical trajectories, such as the one instantiated with the history of the construction state in Korea, should be understood as arising and evolving through the processes of interconnection and interdependence. In other words, seemingly local particularities are always produced and evolve in interconnection with other places and are dynamically integrated into a global capitalist matrix that is ever changing (Palat 1999).

Second, and in a related vein, I also situate pro-greenbelt protests against the greenbelt deregulation that emerged in the late 1990s, as impelled and informed by the contradictions of the locally particular politico-spatiality of Korea's capitalism – that is, the contradictions of the construction state that were amplified by the neoliberalization of space in recent years. In examining these protests, I will refer to Baik's (2013) notion of "core location" – a location of multiple marginalities and the politics of resistance – an investigation of which unveils layered power relations and inequalities. I take the greenbelt in Korea to be just such a core location. And, by examining the contentious history of protests against greenbelt deregulation, I engage with Baik's ideas of "critical self-reflection" (*seongchal*), "praxis" (*silcheon*), and "communicative connection" (*sotong*), and discuss the ways in which pro-greenbelt activism constitutes one node within the broader trans-local "counter-topography" (Katz 2001) of anti-capitalist resistance politics. As many feminist scholars have argued (e.g., hooks 1986; Katz 2001; Mohanty 2003), and as Baik (2013) has also implied, individual sites of resistance are loci of reflexivity of the universal as well as particular conditions of living and life, and, further, a basis of common ground, connections, and the "deep solidarity" (Mohanty 2003, 225) that can be forged among different anti-capitalist movements across borders. In this sense, learning about the pro-greenbelt activism that emerged in 1990s Korea should be registered as one of many efforts "to theorize experience, agency and justice from a more cross-cultural lens" (Mohanty 2003, 244).

Understanding the Local

A recent body of work has raised concerns regarding how most political economic literature in urban studies has failed to represent the multiple modalities of urbanization in non-Western countries. This critical work has problematized the practice of understanding non-Western cities through paradigmatic Western urban theories, such as the "global cities" paradigm and others rooted in the context of Western neoliberal urbanization, and has sought to retheorize the urban from a postcolonial angle (Robinson 2006; Shatkin 2007; Roy 2009). Inspired by a range of postcolonialist theories, including that of Chakrabarty (2000), these scholars have contended that the categories and concepts that seem to have a universal appeal are in actuality provincial, related to the particular circumstance of the emergence of European capitalist modernity. In particular, they have sought to demonstrate the ways in which urbanization in the non-Western world has variegated foundations and evolving patterns, co-determined by multiple local and extra-local factors (see, e.g., Buckley and Hanieh 2014). This scholarship shares an interest in uncovering the different combinations of determination in places, and, based on these, strives for theoretical heterogeneity.

The postcolonialist mandate for decolonizing knowledge production has shaped my interest in the construction state and greenbelt deregulation in Korea. In particular, Chen's (2010) rejection of Eurocentric world historiography and his call for multiplying reference points among Asian subalterns has helped me think through the case that I examine in this chapter (for more details on Chen's arguments, see the introduction to this volume). In contrast to the non-contextualized picture of, say, neoliberal urbanization that is often found in renditions of the global convergence thesis, I am convinced, that urban studies both for Western and non-Western cities should take seriously the path-dependent processes and contextual factors at work in particular places and should carry out more nuanced research. However, it would also be inadequate to analyse urbanization in any particular place in isolation from broader political economic processes that are increasingly becoming the global "rule regime" (Brenner, Peck, and Theodore 2010). In my view, and as many others have also argued, a study of a locality should entail an interrogation of how it becomes a site of *articulations* and *co-determinations* of different forces and processes that have shaped people's experiences of their lifeworlds and their struggles against structural processes, and how local social formations are articulated by, and further integrated into, the broader capitalist world-system through historically and geographically contingent and complex forms (Palat 1999; Hart 2006).

The concept of articulation requires clarification, though. According to Gupta and Ferguson (1992, 8), the concept is often loaded with the assumption that a place is in a "primeval state of autonomy," which is presumed to be transformed and violated by external forces. They argue that, instead of "taking a pre-existing, localized 'community' as a given starting point" (ibid.), as has been implied in the concept of articulation, it is necessary to examine how each place has *always* been formed out of interconnected processes (that are often in tension with each other) originating from, and moving through and between, different places. The differences that each locale possesses are never pristine or uncontaminated, but rather are produced through socio-spatial forms of interconnection and interdependence. Therefore, the fault line between what is purely local and what is extra-local is hard to discern. While Gupta and Ferguson develop the notion of articulation to explain culture and place, a similar point has continuously been made by Marxists, too. Marxists have shown capitalism's innate tendency toward competition for absolute and relative surplus value for accumulation's sake and the consequent expansion of the system to wider geographical and social realms. Harvey (1989) shows that such a tendency results in "time-space compression," where different places in the world are intimately integrated and concatenated through increasingly advanced transportation and communication technologies.[2] Leon Trotsky's theory of "uneven and combined development," while it acknowledges the different, multilinear historical progression of different places,

---

2 I thank Kyle Gibson for pointing out this issue to me.

also illuminates how these places are interactive with each other and integrated into the capitalist system via trade and capitalist overseas investment (Anievas and Nişancıoğlu 2017).

Such points about articulation enlighten those of us who study Korea to bear in mind that a historically informed and contextually grounded study of the country should not lead us to a preoccupation with the supposedly incompatible particularity of Korea. Rather, they should compel us to explore the historically sedimented particularity of Korea that has been formed through interconnected geopolitical and capitalist processes, as well as transnational flows of ideas and forms of praxis between different places. In this sense, the tendency of certain post-colonialists to reject the universalization of the capitalist system – the latest incarnation of which is neoliberal capitalism – cannot properly capture the totality of life in any place. In some studies, the particular is misleadingly equated with the local and is analysed as being incompatible with the universal and abstract (Sayer 1991; Hart 2006, 996; Orzeck and Hae forthcoming). This is an analytical fallacy, which is empirically unsustainable, and we need to take seriously the question of the universal in our study of locality. We should also remember, however, that the universal is not an "epistemological given." Instead, as Chen (2010, 245) reminds us, it should be understood as a horizon constructed through locally based grounded knowledge.

Again, it is important to theorize local differences in their mutually constitutive relation to the broader macro structure and the increasingly universalizing capitalist market compulsion (which takes specific historical forms). And, as is mentioned earlier in this chapter, an attention to local differences has been one of the under-explored dimensions of some political economic literature in urban studies. Another subject that these schools of thought have also under-privileged in their theoretical endeavours has been the everyday political struggles of people. According to Ruddick and her colleagues (2017), the planetary urbanization thesis (Brenner and Schmid 2015), for example, does not consider struggles and practices of people as the key component and generative forces constituting the social ontology of the urban, and its optic stays at the level of the abstract.[3] Theories that attend to struggles

3 First conceived by Henri Lefebvre ([1970] 2003), the idea of "planetary urbanization" was recently revitalized through the work of Brenner and Schmid (2015). The latter argued that capitalist urbanization – in particular, neoliberal urbanization – now cuts across different kinds of spaces (including agricultural and wilderness spaces) and takes a planetary form. This, according to them, compels urbanists to revise inherited categories of, and the binary between, the urban and non-urban.

and take praxis seriously, such as Baik's idea of core location, can, therefore, provide a corrective to such a tendency among some political economist works in urban studies, even though Baik's concept is not free of limits and political ambiguities, as is discussed in the introduction to this volume. For the episteme of Baik's core location starts its analytics from the marginalized places and people and their resistant actions against complex relations of power.

At the heart of Baik's concept of core location is the triad of critical self-reflection, praxis, and communicative connection, which he posits as the sources through which resistant forces challenge and eventually transform formidable material structures that constantly generate disparity, dispossession, and marginalization. He emphasizes the importance of mutually understanding different place-based ideas and praxis, and connecting them across different networks and scales. This approach resonates with many feminist scholars' arguments that critical activist scholarship should seek to contribute to increasingly globalizing organizational struggles and oppositional politics, while at the same time also pursuing non-colonizing and decolonizing scholarship. Non-colonizing and decolonizing scholarship, Baik contends, starts from grounded knowledge of places and the particular conditions in which these places are situated and that shape their peripheral states. This echoes Mohanty's (2003, 223) call to scholars to recognize that "the particular is often universally significant," while at the same time to not try to subsume the particular within the universal.

These ideas by Baik and Mohanty are the framework through which I want to read the story of the pro-greenbelt activism that emerged in the late 1990s in Korea. In the next two sections, I attempt to provide an account of the historical and conjunctural processes that triggered the counter-movement by pro-greenbelt activists against the state's greenbelt deregulation. The processes that are described below are embedded in specific local histories of Korea, but they are also enmeshed in wider relational capitalist, geopolitical processes.

The "Construction State": The Particular Capitalist Political Form in Korea

The particular spatio-political assemblage within which I locate the story of pro-greenbelt activists in Korea is the "construction state." That term expresses a capitalist regime of accumulation and regulative modes developed in Korea in relation to the country's particular position in, and interconnection with, broader geopolitical constellations that were formed during the Cold War period.

The authoritarian and military "developmental states" of the period from the 1960s to the 1980s in Korea facilitated rapid industrialization and economic growth, through which these regimes, which seized power via coups, sought to maintain their political legitimacy (Cho 2013, 269). This economic growth proceeded through a hard-line offensive against the principle of democracy in the political, economic, and social spheres. The blank slate granted to the military regimes also stemmed from Cold War geopolitics in which the Pacific ruling-class alliance, especially the US government, supported the Korean military regime against the East Asian communist bloc that arose around it (Glassman 2016). Since the late 1980s, however, the processes of democratization, neoliberalization, and administrative decentralization gradually destabilized the previous years' authoritarian developmental statist system, and mixing with (and often contradicting) one another, these new processes gave rise to new political configurations, which some have characterized as "post-developmental" (Doucette 2016).

Urbanization has experienced similar transitions. Neoliberal urbanization emerged in the late 1980s and quickened after the economic crisis of the late 1990s (Choi 2012). During the developmental statist period, the state virtually monopolized spatial planning. By contrast, each regime in the post-developmental period implemented an inventory of urban policies that offered more initiatives for private participation, giving in to the development industry's persistent demands for deregulation in matters of land use and development. In the meantime, new forms of popular democracy and civic engagement have exploded since the 1990s along with political democratization in the country. Environmental and other dissident groups have entered the political arena, raising new claims to a set of rights and entitlements that had been denied them during the authoritarian developmental state era (Y. Lee and Shin 2012). At the same time, with administrative decentralization, local governments have surfaced as key political agents.

Despite these shifts in urban governance, some observers have noted that the recent transformations in urbanization processes have taken a "path-dependent" character, with developmental statism still lingering, a view that leads some commentators to label the current regime shift as "neo-developmental" (Choi 2012; Cho 2013) rather than "post-developmental." The central government still exercises significant institutional power in matters related to urban/land development, and it continues to act as one of the primary developers (Y. Lee and Shin 2012; Choi 2012; also see Oh's chapter in this volume). In particular, urbanists have concurred that the "construction state" is one of the core features of the politico-institutional structure that has survived the transition from the developmental to the post-/neo-developmental period.

The term "construction state" originally referred to a specific Japanese economic structure (McCormack 1995), but Korean scholars and activists maintain that the Korean counterpart operates as more or less the same machine (Hong 2011; Cho 2013). The construction state in Korea (as well as in Japan) refers to the system in which the proportion of the construction sector is substantive in the country's GDP, and the "iron triangle" of the "construction complex" – that is, companies-bureaucrats-politicians – exercises a powerful hegemony in the state's affairs (McCormack 1995).[4] In Korea, it was not until recent years that the term became a scholarly and activist concept (Hong 2011; Cho 2013). Scholars and activists have since used the term to illuminate the current state and contradictions of urbanization in Korea and to mobilize activist groups to resist the construction machine of the country. While the rising importance of real estate and construction within national economies across the world has become universal, as neoliberal regimes have proliferated since the early 1980s, in Korea it by far predates that period: the construction state mechanism has existed as a territorially embedded capitalist, politico-institutional development since the 1960s.

The construction state is a unique modern capitalist urban modality, and "local" and "extra-local" forces have coalesced to constitute it over the past few decades. The origin of the construction state in Korea can be traced to the nation-building drive and recovery efforts following its independence from Japan as well as the Korean War, both of which drove the state to prioritize the (re)construction of the physical infrastructure (Park 2011, 209). At the same time, however, as Glassman and Choi (2014) demonstrate, the US military offshore procurement contracting that Korean (proto-)*chaebols*[5] engaged in during the Vietnam War bestowed upon them, especially Hyundai, a pivotal opportunity for technological learning and the upgrading of engineering and management skills in the fields of construction, which precipitated the advent of the modern construction industry and the construction state in the country. Glassman and Choi develop this argument to debunk the "national–territorial" and state-centric frames of the so-called neo-Weberian accounts of the Korean developmental state. They highlight as an alternative the significant enmeshment of both the developmental

4 Politicians in the construction complex include those operating within national, regional, and local governments and parliamentary institutions.

5 *Chaebol* refers to a family-owned, large business conglomerate that has been formed and developed via governmental supports since the 1960s in Korea.

and the construction state within Cold War geopolitics and, in particular, the US military-industrial complex. This argument shows how the Korean capitalist political economy that is often referred to as developmental statism was not a feature generated solely by local agents and mechanisms – that is, the strong state and rational-planning bureaucracy, as neo-Weberians have argued – but the outcome of transpacific interconnections, echoing a point discussed earlier in this chapter in relation to Gupta and Ferguson's work (1992) (see also Glassman 2016).

During the authoritarian Park Chung-Hee regime (1963–79) – the key developmental statist period – civil engineering and public works became a central feature of the Korean political economy, as the export-oriented, manufacturing-based economy required massive investment in productive and reproductive infrastructure. Large land and housing development projects were virtually monopolized by the public sector. In this period, the machinery of a construction bureaucracy (*togeon-gwallyo*) – composed of public corporations established for property and land development and the ministry in charge of national construction affairs – became gigantic and powerful.[6] These key elements have been the linchpin of the construction complex and have been one of the sources of crony capitalism, exercising (often illegitimate) favouritism toward a privileged few capitalists, as well as landed and propertied classes. They frequently implemented policies that would enhance the popularity of the specific regime they served (Hong 2011). Massive public investments in land and housing developments by the public sector during this period were central in consolidating the country into a construction state.

The power of the construction state continued into the 1980s. In this period, the two military regimes continued with their crackdowns on anti-government dissidents. At the same time, they proceeded with several construction projects – most representatively, the "2 million housing" construction project" in the form of massive apartment complexes and the development of suburban towns surrounding Seoul. These massive constructions, planned and implemented mostly by public corporations, were meant to appease discontented urban inhabitants in Seoul. The developmentalist capitalist regime also began experiencing an over-accumulation crisis during this period, and the regimes sought

6 The name of the ministry has changed over time: from 1962 to 1994, it was the Ministry of Construction; from 1994 to 2008, the Ministry of Construction and Transportation (MOCT); and from 2008 until 2013, the Ministry of Land, Transport and Maritime Affairs (MOLTMA). Since 2013, it has been the Ministry of Land, Infrastructure and Transport (MOLIT).

to tackle this by channelling surplus capital into the real estate sector (Shin 2009; Harvey 1982). In the midst of implementing this array of construction projects in the 1980s, the construction complex was further consolidated.

From the 1990s onwards, while democratization could have potentially tamed the construction state drive, the neoliberal policy platform pursued by political regimes in this period – both left and right – actually strengthened the power of the construction state. Deregulation became a political mantra, especially after the financial crisis in the late 1990s. That crisis resulted in an International Monetary Fund (IMF) bailout of the country and a deregulatory wave that arose from the structural adjustment programs imposed by the IMF during the Kim Dae-Jung regime (1998–2003). Construction projects proceeded apace during this period, as more capital moved to the deregulated and market-oriented real estate sector. Despite this deregulatory wave, the state's presence in urban development was also reinforced, albeit in a modified form, as the state sought to enhance its national competitiveness through a range of spatial developments (Cho 2013, 271). In the early 1990s, administrative decentralization commenced, which decentred even further the former developmental state. Yet this did not lessen the power of the construction state. Each region became embroiled in inter-regional competition over central government funding as well as private mobile capital (both foreign and domestic), and launched into a stream of construction projects under the leadership of local politicians to enhance local competitiveness (Park 2011, 215; on this trend, see also the chapters by Eom, Jeong, Oh, and Park in this volume).

Although the rhetoric that the construction complex used to justify an unceasing construction wave in the so-called neo-developmental period from the 1990s onward became more reformist in tone, many contend that the construction state's paradigmatic logics have remained intact (Cho 2013; Hong 2011). That is, private interests are still favoured by the state; the processes of planning and development have been largely undemocratic; and the pernicious impacts of these constructions, such as growing land and housing speculation, have disproportionately affected the economically disenfranchised working-class population. Populations with disposable income continue to seek to profit from their investments in real estate, with speculative subjectivities having become one normalized component of capitalist modernity in the country (Sohn 2008). It is also worth noting that the non-stop implementation of construction projects had been established as the sine qua non for the reproduction and consolidation of the interests of the construction complex. By the early 2000s, the housing provision ratio reached 100 per cent in Korea. But whenever housing shortages became a thorny political issue, the argument advanced by the construction complex has been to "build more," even though the problem has been more the unequal distribution of ownership among the populace. Many have conceived this phenomenon as arising from the calculation of the construction complex to reproduce itself and maintain its hegemony (Hong 2011; Cho 2011, 2013). This also explains the proliferation over the past couple of decades of construction projects with little fiscal feasibility.

This observation does not imply that the ongoing persistence and vigour that social and environmental groups have shown in opposing the construction complex can be discredited. On the contrary, these groups' activism against the construction state has been strikingly vehement, unyielding, and persistent. One instance of such resistance politics was waged by pro-greenbelt activists. In the following section, I examine the process of greenbelt deregulation, which will be followed by an account of the pro-greenbelt struggle of environmentalists and social activists against the construction state machine that executed greenbelt deregulation.

The Greenbelt as Core Location

Korea's greenbelt policy was first drafted and imposed in the Seoul area and thirteen other cities by the Park Chung-Hee regime in 1971.[7] Development restrictions in the form of a greenbelt policy may seem paradoxical, considering the developmentalist mandate of that era. Yet "national security" concerns related to the Cold War geopolitics of the time factored in more heavily in the designation of the greenbelt than did concerns for development control. Indeed, the major purpose of the greenbelt was to secure sites for strategic military action in preparation for potential war with North Korea (Chang 2004, 70; Jung 2005, 126). Nonetheless, the regime's designation of greenbelts was also meant to curb urban sprawl, increasing real estate speculation in major cities experiencing rapid population growth, and the contamination of air and sources of drinking water.

The greenbelt designations were driven by administrative convenience and a dictatorial, centralized planning mechanism. They were undertaken by technocrats who did not conduct thorough land-use

7 The area of greenbelts in total was 5,397 km^2 and covered 5.4 per cent of the total national land (MOLTMA 2011, 22).

surveys or seek public input, and who drew the boundaries of greenbelts based only on air and topographical maps. This was a typical example of the sort of undemocratic decision-making processes conducted behind closed doors by the authoritarian developmentalist regime of that time. The designation of greenbelt lands in this way not only caused serious inconveniences to residents in these areas – most of whom were engaged in agricultural activities – but also seriously restricted landowners' rights. This was a period in which the national interest, especially the security interest, was prioritized over individual interests (often including property ownership) by a McCarthyesque dictatorial state machine, so land and property owners did not dare to challenge the state's decisions regarding the greenbelt for fear of harsh retribution.

Since their establishment, the physical boundaries of the greenbelt have barely changed. Yet, given the circumstance that more than 80 per cent of greenbelt lands were privately owned, controversies over the strict greenbelt regulations continued, and in the 1980s the new military regimes permitted minor relaxations of the regulations to modestly improve the convenience and livelihoods of greenbelt residents. The critical change in this period, however, was that the state started to approve the locating of public buildings and amenities, such as administrative buildings, recreational public parks, educational tourist farms, and sports facilities, in greenbelt lands (Chang 2004, 75; MOLTMA 2011, 148–9, 189). The construction of public buildings and amenities was far from being uncontested, with the media in particular engaging in serious criticism of the government, especially when controversial facilities, such as waste treatment facilities and golf courses, were also permitted to be built (MOLTMA 2011, 152). The approval of these developments then gave rise to increasing discontent among greenbelt landowners and residents, as they perceived – quite correctly – that the greenbelt rules had been strictly enforced on them alone (Kim and Kim 2008, 46).

This particular evolution of greenbelt development signified to many that the state was not actually interested in protecting the greenbelt. If anything, many contended that the greenbelt became a reservoir for contemporary public development (Chang 2004, 75). Furthermore, the increased access to greenbelt lands by public institutions together with a wave of (limited) relaxations of greenbelt regulations raised the general expectation that the greenbelt would soon be deregulated and opened to private forms of development. It became commonplace by this time for the affluent class to buy up greenbelt lands, especially in the Seoul Metropolitan Region (SMR), from the original owners and to start to build luxury houses and restaurants on these lands (MOLTMA 2011, 170, 171). They frequently bribed public servants in charge of approving use changes to greenbelt lands (176–7). According to one media report published on 27 June 1990, a public officer commented that "the preservation of greenbelts is dependent upon the state's will to curtail the power of this affluent class" (176). The price of some parts of the greenbelt surrounding the Gangnam area of Seoul – which is the most expensive district in the city – experienced an upswing, and speculation over these lands ensued.

In the meantime, from the late 1980s to early 1990s, suburbs in the SMR were developed as a massive residential area via new town developments – such as the 2 million housing construction project, mentioned above, which was a landmark feat of the construction complex. Lands there were subject to frantic speculation, and landowners enjoyed hefty profits as a result. This further augmented the discontent among greenbelt landowners in the SMR whose lands were not part of speculative frenzy, owing to development restrictions. Speculation had by then been established as one of the major means of upward mobility in the country and one key component of the construction state's operating mechanism. Greenbelt landowners started to mobilize themselves, and by the mid-1990s they became active political participants. The continued political democratization of the country further aided this process, emboldening these landowners to speak out. They claimed that it was unfair to impose the social costs of preserving greenbelts on greenbelt landowners. While this advocacy was an understandable move by owners whose property rights had been severely infringed upon, not all who embarked on the political action were victims of the previous system. The class composition of greenbelt landowners had changed during the intervening years. According to a survey conducted by the Ministry of Construction and Transportation in 1993, the percentage of greenbelt landowners who had lived in greenbelt from the time of its designation was down to 45 per cent, while that of the greenbelt landowners who lived outside greenbelts – that is, rich landowners and/or speculators – was 46.3 per cent (MOLTMA 2011, 192). Many suspected that the latter group, rather than the former, was spearheading the campaign to completely repeal greenbelt regulations.

Meanwhile, in the mid- to late 1990s, two critical changes – one geopolitical and the other institutional – steered the state in the direction of relaxing greenbelt regulations. First, the idea of relaxing these regulations was aided by the easing of the Cold War tensions that had initially given rise to greenbelt designation in the 1970s (Chang 2004, 77, 84). Second, with administrative decentralization and the commencement of a local autonomy system in the mid-1990s, which was an outcome of broader political democratization, the central government was pressured by local governments to ease greenbelt regulations (MOLTMA 2011, 141). Local governments started to cut back, and even boycotted allocating, budgets for the enforcement of greenbelt regulations, contending that they should not be mandated to earmark funds for a task that was essentially a breach of their constituents' basic property rights (Jung 2005, 128).

Within this context, the 1997 presidential election offered a political opening to greenbelt landowners, and they pledged the support of their 740,000 votes to the candidate who would execute the repeal of the greenbelt policy. The greenbelt landowner organizations supported the centre-left candidate Kim Dae-Jung – one of the most prominent political opponents of the military governments in previous decades – who promised to execute a drastic deregulation of greenbelt policies. Kim saw such regulation as a regrettable legacy of the previous president Park's military regime, which he himself had fought against. Ultimately elected as the first president from an oppositional party in the history of South Korea, Kim commenced the reform of greenbelt regulations by declaring a principle of "releasing areas that need to be released [meaning 'impaired areas'] while tightening areas that need to be tightened" (*pul geoseun pulgo mukkeul geoseun mungneunda*) (Chang 2004, 85).

Greenbelt deregulation was also in step with the broader deregulatory initiatives implemented by the Kim regime following the IMF bailout of the country in the late 1990s. Perceiving the greenbelt as a massive land stock that showed great potential, but was still awaiting full development, the key members of the construction complex – state actors, experts in government-owned public corporations, urban practitioners, and allied development capitalists – believed that reviving the real estate and land market through the deregulation of the greenbelt would help the economic recovery in the wake of the crisis. The mandate to follow this global deregulatory trend was strong among state officials and technocrats. Additionally, as Seoul was struggling with a chronic shortage of housing, urban experts as well as politicians insisted that greenbelts around Seoul should be released so that more housing could be built there. The importance of the greenbelt as a key site of ecological protection was relegated to the margins in the midst of these political and economic calculations, and the construction state assemblage rose as the mode of governing the greenbelt, now informed by the neoliberal ethos that was increasingly gaining currency in the state's urban policies.

Pro-greenbelt Activism: Resisting the Capitalist State and Its Spatial Modalities

The landscape of resistance politics in Korea was dominated by the labour movement until the end of the 1980s; but from the 1990s onward, environmental activism together with other new forms of activism surfaced as a discernible force in the country's political landscape. Such activism was simultaneously an offspring and facilitator of the broad democratization process of the time, which contributed to a post-developmental political constellation in the country. With the state's move to deregulate the greenbelt, a movement led by a pro-greenbelt coalition emerged in the late 1990s. The coalition stressed the urgency of protecting greenbelts from the state's imminent act of deregulation. Its campaign asserted that the greenbelt is ecologically important and that its preservation should be considered a means of enhancing the public good (Jung 2005; Chang 2004, 87). Referring to greenbelts as a "life-belt" (*saengmyeong-belteu*) (Bae and Lee 1999), the activists in this coalition proclaimed that these areas help preserve biodiversity around major cities, prevent flooding, and attenuate climate change (MOLTMA 2011, 338). Furthermore, they argued that, without a greenbelt, major cities would be left with less green space and with more intensified urban sprawl, and that, with greenbelt deregulation, cities, especially ones surrounding Seoul, would experience greater land speculation. They demanded that the state pay overdue compensation to greenbelt landowners for their long-lost rights to their properties, but leave the greenbelt intact.

Activists also contended that illegal impairments to greenbelts had been made by the people who bought greenbelt lands for speculative purposes (as well as by public institutions) in the previous decade, and, thus, that releasing greenbelt lands that had experienced impairments would benefit only the culprits who had degraded the land. What the state was doing, they argued, was tantamount to a denial of the real history of greenbelt degradation. Critics and activists further maintained that greenbelt deregulation was the key operating mechanism of the construction state – the logic of constructing more housing to solve the housing problem – being emphatically put forward by the construction complex, which they called the "construction mafia" (Choi 1998, 23; Hong 2011). Calls to "resist the construction state" became the lingua franca that connected different social activist and environmentalist groups into a pro-greenbelt activist coalition.

As the cries of these pro-greenbelt organizations were increasingly gaining public attention, the MOCT decided to have pro-greenbelt activists represented in the Committee for the Revision of the Greenbelt Policy (CRGP, Geurinbelteu Jedogaeseon Hyeobuihoe), a counselling body for the process of drafting a new greenbelt law (Jung 2005, 129). Founded in May 1998 and composed of a range of "civil society" actors as well as public officials, it was created to achieve what they called a "social consensus" regarding greenbelt deregulation. However, only two members from the pro-greenbelt coalition were permitted on the twenty-three-member committee, which meant a serious power imbalance in its decision-making process (Chang 2004, 66).

This imbalance, of course, eventually had its repercussions. In the seven months following its formation, CRGP announced a first draft of the plan for greenbelt reform. Not long before the announcement, the Korean Constitutional Court declared that the expropriation of a person's land and property rights through greenbelt regulation would be unconstitutional if it were not coupled with proper compensation by the state to the affected people. As a result, the state felt it even more pressing to abolish greenbelt regulations (MOLTMA 2011, 247). Overall, the reform suggested by CRGP was to completely remove greenbelt regulations from thirteen small- to medium-sized cities and, in the case of major metropolitan regions such as SMR, to abolish the greenbelt category only in areas with lower environmental values. This plan was a shock to pro-greenbelt activists, as many small- to medium-sized cities included lands that, according to several environmental impact studies, should be protected because of their ecological sensitivity or the water sources in them (Yang 1999; Bae and Lee 1999, 82, 83). With this announcement of the tentative plan for revising greenbelt laws, local governments of these cities started to announce plans to build casinos, leisure facilities, and/or golf courses on the greenbelt lands that were slated for deregulation (Bae and Lee 1999, 86).

Consequently, pro-greenbelt activists scaled up their acts of resistance. The two pro-greenbelt activist members on CRGP threatened to resign their posts. A number of pro-greenbelt activists attended the public hearings that followed the announcement of the deregulation plan and vehemently raised their objections to it. They also recruited more sympathetic activist organizations and converted themselves into an umbrella coalition called the People's Action for Saving Greenbelts (PASG, Geurinbelteu Salligi Gungmin Haengdong). The PASG launched a repertoire of combative direct actions, rallies, marches, forums, and newspaper advertisements, which helped it lead a good number of media outlets over to the pro-greenbelt side (Jung 2005, 124;

Bengston and Youn 2006, 69).[8] Greenbelt landowners often interrupted PASG-organized forums and rallies, as was illustrated at the beginning of this chapter, and frequently made threatening calls to the PASG's key activists (Jung 2005, 125).

The PASG's investigations revealed that six of the thirty members of the MOCT committee of the National Assembly were current landowners in greenbelts destined for deregulation (Bengston and Youn 2006, 73), revealing that these members had a personal stake in the issue. At the same time, the PASG organized "five working level talks" with MOCT officials and other state bureaucrats (Jung 2005, 129). Importantly, the Ministry of Environment (MOE)[9] was on the side of greenbelt preservation and counterbalanced the construction complex (at the centre of which stood the MOCT and the Blue House) that was driving greenbelt deregulation. Activists expected this splintering within the state apparatus to further empower the pro-greenbelt movement (Chang 2004, 80).

The unceasing campaigning waged by the PASG forced the MOCT to make concessions, agreeing to the pro-greenbelt coalition's demand for a neutral third party to review the tentative plan. At the end of 1998, the MOCT commissioned the Town and County Planning Association (TCPA) from the United Kingdom to assess the tentative plan, review the overall state of greenbelts in Korea, and propose some suggestions for greenbelt governance. Following the report from the TCPA, the MOCT announced a revised and final greenbelt policy in July 1999. Although, as discussed below, it did incorporate one crucial demand from the PASG and the TCPA, the final law largely bypassed a number of key demands made by the PASG and preserved much of the content of the first draft's pro-deregulation position. The final plan was also made public by the MOCT before it acquired consensus from the CRGP (Chang 2004, 86n15).

The PASG resisted the new plan by orchestrating protest spectacles in various parts of the country. Some members shaved their heads in public, carried out overnight sit-ins, and staged hunger strikes. They also held a press conference, collected a petition signed by a million people to impeach the minister of the MOCT, and sued the MOCT in

8 Some scholars and activists also began to discuss the possibility of applying the National Trust movement to greenbelts (Bae and Lee 1999; Oh 2000). The National Trust is a campaign in which interested citizens purchase historic sites or ecologically sensitive areas to establish them as commons and preserve them.

9 The MOE was established in 1994.

the Supreme Court for its violation of environmental rights (Chang 2004, 86). Individual activists were carrying out campaigning in major squares and plazas of Seoul under the hot summer sun and torrential summer rains. By this time, the number of participating organizations – mostly social activist and environmentalist groups – in the PASG amounted to 247 (Jung 2005, 124). Eventually, the MOCT scaled back some of its deregulation plans and reluctantly agreed to reduce the number of deregulated greenbelt lands from 113 to 94 (Chang 2004, 88n20). Ultimately, though, the activist coalition could not stop the powerful current of greenbelt deregulation pushed forward by the developmentalist construction complex, which was buttressed by greenbelt landowners and liberal economists and urban experts. The MOE – a crucial source of counter-pressure to the construction complex – was eventually relegated to only a passive (consulting) role in the process of revising the greenbelt laws by the MOCT (Chang 2004, 87). The pro-greenbelt movement was unyielding and persistent, to be sure, but the balance of political power was deeply asymmetrical.

In retrospect, PASG activists realized that they should have organized their movements at the local scale in order to "construct grievances at the grassroots level" instead of using the abstract language of "ecological preservation" to appeal to the general public (Jung 2005, 129). Some PASG activists also concluded that the failure of the movement could be ascribed to the misguided belief at its birth that talks with the MOCT and working within the CRGP would be an effective strategy (H. Lee 1999). They claimed that the government's placement of opponents of greenbelt deregulation on CRGP was at best a public relations stunt. The PASG became disorganized shortly after the new greenbelt law was passed, and activists from various groups started to scale down their activism to local chapters (Jung 2005, 130). Activist organizations located outside Seoul, however, often suffer from a lack of expert cohorts and full-time activists as well as from under-funding, and this has impeded these local chapters from actively monitoring urban developments in deregulated greenbelts (ibid.). This consideration makes it difficult to argue in general that administrative decentralization would have led to local democratization, as the mandate of local competitiveness pronounced by local political elites, capitalists, and the landed class – members of the construction complex – prevails in the absence of well-resourced countervailing social activist forces.

Despite this defeat, pro-greenbelt activism was not in vain. One of the PASG demands was that the government establish a long-term, principled metropolitan regional planning mechanism (which had not existed in the Korean planning system up to that point) and situate the process of greenbelt deregulation within it. Such a mechanism was intended to prevent unplanned and haphazard development (*nan-gaebal*) in the greenbelt, as well as to conform greenbelt deregulation and development to the comprehensive large-scale planning (Jung 2005, 132; Chang 2004, 80). This call from the PASG was buttressed by the TCPA's recommendation that the metropolitan regional planning body be established before the greenbelt was deregulated. This demand was eventually incorporated into the final draft of the revised greenbelt law.

The PASG's achievements did not end there; it also had a crucial impact on later environmentalist movements. The PASG, as one of the earlier large-scale environmentalist groups, set the example of why and how to (or how not to) fight the construction state. Additionally, according to one interviewee in Jung's research, the greatest achievement of the PASG was the "transformation of public discourses on greenbelts from the issue of public taking of private assets to environmental preservation" (2005, 129). The activism powerfully recreated the image and importance of greenbelts as an environmental reservoir that was threatened by the construction state machine, rather than as a symbol of dispossession associated with the previous authoritarian state. For example, in a survey conducted by the MOCT in 1999, randomly selected groups of citizens and planning experts expressed that they preferred the preservation of the greenbelt with minimal changes, whereas the majority of the residents of the greenbelt preferred its deregulation (125). If a yearning for an ecologically friendly quality of life in order to combat the rapid pillaging of the urban natural environment can be understood to stem from middle-class cultural proclivities, the survey actually showed that a wider range of different classes sympathized and concurred with the key claims made by pro-greenbelt movements, rather than looking upon greenbelts as barriers to lucrative urbanization (Chang 2004, 76–7, 79). Pro-greenbelt activism appealed to this sensitivity and elevated this yearning to a set of concrete rights claims.

But popular consciousness was also paradoxical. For example, in a survey conducted by the MOCT, 36.8 per cent of respondents were opposed to the tax increases necessary to preserve the greenbelt (Chang 2004, 84). As environmental consciousness among the masses grew only slowly and was frequently interrupted by people's short-term economic self-interest, it was hard for environmentalist groups to receive sustained support from the middle- and working-class constituencies (ibid.).

Conclusion

Since the revision of greenbelt laws in the late 1990s, the state has continued to increase the size and scope of the deregulation of the greenbelt, while also building apartment complexes in deregulated areas. Such development was often carried out by breaking deals that the state had reached with the pro-greenbelt activist coalition during the greenbelt law revisions at the end of the 1990s. The construction state endured throughout this period, constantly seeking to quiet dissident environmentalist voices. However, developments on greenbelts in the first two decades of this century also generated combative counter-actions from environmental activists in multiple pockets of the greenbelt sites. In some areas, activists established alliances with greenbelt residents who had grown disillusioned with the way the greenbelt was being developed. The state was forced to accommodate some demands from these dissidents, and in some greenbelt areas it had to cancel deregulation and development plans entirely. One lesson to derive from this story is that the central planks of resistance politics should be tireless organizing, persistent political engagement, and the building of solidarity among different actors.

The history of greenbelt deregulation reveals the greenbelt as a core location – that is, a place where the ecological considerations associated with it are relegated to the margins by the players in the construction complex. The fight waged by pro-greenbelt activists was driven by the territorially embedded, specific capitalist mechanism of the construction state, which has also been formed and consolidated through interconnection, articulation, and integration within global capitalist processes. Greenbelt activism in Korea, therefore, resonated with a range of resistant forces, movements, and campaigns that have emerged across the world to oppose capitalist states, as well as real estate and landed capital that have turned natural and built environment into sites of speculation. Studying greenbelts and greenbelt activism in this way offers "one route toward an understanding of world history" (Chen 2010, 253) and provides a method through which to understand the contours of a trans-local topography of resistance. One way that academics studying one specific place can contribute to this trans-local movement is to chart the "loops of codetermination and coevolution" (Buckley and Hanieh 2014, 158) of different forces that shape the actually existing social world in individual sites, identify old and new forms of domination and subordination that are also connected to the broader global capitalist system, and point to the cracks, ruptures, and contradictions of systems that may open up political spaces for on-the-ground dissident politics. That is, the task at hand for researchers is to examine ongoing articulations and co-determinations of the different forces and processes at work in a given place, while simultaneously being reflexive in terms of universal(ized) categories, imaginaries, and optics coming from Western paradigms, when researchers seek to conceptualize place-based processes.

The achievements and defeats of pro-greenbelt activism, its engagements within and outside the prescribed political space, its resilience and incredible commitments to ongoing struggles, and even the mundane rallying cries its proponents chanted, provide clues that illuminate the state of contemporary resistance politics and the possibilities of transformation that they represent. As Chen (2010) stresses, the study of a place necessarily transcends that place. Understanding struggles over greenbelts in Korea is a step toward the imagining of common ground and an informed and reflexive solidarity between different movements against exploitative, dispossessive capitalism in which the capitalist state is a crucial entity. Studying the struggles over the greenbelt in Korea can, therefore, help us "foreground not just the connections of domination but those of struggle and resistance" (Mohanty 2003, 243).

REFERENCES

Anievas, Alexander, and Kerem Nişancıoğlu. 2017. "Limits of the Universal: The Promises and Pitfalls of Postcolonial Theory and Its Critique." *Historical Materialism* 25 (3): 36–75.

Bae, Jeonghwan, and Geunhang Lee. 1999. "Donghyanggwa Jaengjeom – Hwan-gyong-undongui Saeroun Mosaek: Geurinbelteu, National Trust Woondong [Trends and Issues – In Search of New Environmental Movement: Greenbelt National Trust Movement]." *Environment and Life* 21: 80–93.

Baik, Young-seo. 2013. *Haeksimhyeonjang-eseo Dong-asiareul Dasi Mutda: Gongsaengsahoereul Wihan Silcheon-gwaje* [Rethinking East Asia in Core Locations: A Task for a Co-Habitation]. Paju, KR: Changbi.

Bengston, David N., and Yeo-Chang Youn. 2006. "Urban Containment Policies and the Protection of Natural Areas: The Case of Seoul's Greenbelt." *Ecology and Society* 11 (1): 65–79.

Brenner, Neil, Jamie Peck, and Nik Theodore. 2010. "Variegated Neoliberalization: Geographies, Modalities, Pathways." *Global Networks* 10 (2): 182–222.

Brenner, Neil, and Christian Schmid. 2015. "Towards a New Epistemology of the Urban?" *City* 19 (2/3): 151–82.

Buckley, Michelle, and Adam Hanieh. 2014. "Diversification by Urbanization: Tracing the Property-Finance Nexus in Dubai and the Gulf." *International Journal of Urban and Regional Research* 38 (1): 155–75.

Chakrabarty, Dipesh. 2000. *Provincializing Europe: Postcolonial Thought and Historical Difference*. Princeton, NJ: Princeton University Press.

Chang, Se-Hoon. 2004. "Geurinbelteuui Jeongchisahoehak: Geurinbelteu Haeje-ui Sahoe Donghageul Jungsimeuro [The Political Economy of the Greenbelt: The Social Dynamics Surrounding the Deregulation of Greenbelts]." *Kyongjewa Sahoe* [Economy and Society] 63: 65–99.

Chen, Kuan-Hsing. 2010. *Asia as Method: Toward Deimperialization*. Durham, NC: Duke University Press.

Cho, Myung-Rae. 2011. "Gonggeupmanneungjuuireul Beoryeoya Jutaekjeongchege Sanda!" [Supply Centrism Should Be Eliminated for a Better Housing Policy!]. *Kyungki Daily*, 11 July.

Cho, Myung-Rae. 2013. *Noksaektogeonjuuiwa Hwangyong-wigi* [Greenwashing Developmentalism and Environmental Crisis]. Paju, KR: Hanul Academy.

Choi, Byung-Doo. 1998. "Budongsan Gyuje Wanhwa Mit Geurinbelteu Jojeong-ui Han-gyewa Daean" [The Limitations and Alternatives to the Deregulation of Real Estate and Greenbelt Readjustment]. *Journal of Urban Studies* 4: 13–31.

Choi, Byung-Doo. 2012. "Developmental Neoliberalism and Hybridity of the Urban Policy of South Korea." In *Locating Neoliberalism in East Asia: Neoliberalizing Spaces in Developmental States*, edited by Bae-Gyoon Park, Richard Child Hill, and Asato Saito, 86–113. Hoboken, NJ: Wiley-Blackwell.

Doucette, Jamie. 2016. "The Post-Developmental State: Economic and Social Changes since 1997." In *Routledge Handbook of Modern Korean History*, edited by Michael J. Seth, 343–56. London: Routledge/Taylor and Francis.

Glassman, Jim. 2016. "Emerging Asias: Transnational Forces, Developmental States, and 'Asian Values'." *Professional Geographer* 68 (2): 322–9.

Glassman, Jim, and Young-Jin Choi. 2014. "The Chaebol and the US Military-Industrial Complex: Cold War Geopolitical Economy and South Korean Industrialization." *Environment and Planning A* 46 (5): 1160–80.

Gupta, Akhil, and James Ferguson. 1992. "Beyond 'Culture': Space, Identity, and the Politics of Difference." *Cultural Anthropology* 7 (1): 6–23.

Hart, Gillian. 2006. "Denaturalizing Dispossession: Critical Ethnography in the Age of Resurgent Imperialism." *Antipode* 38 (5): 977–1004.

Harvey, David. 1982. *The Limits to Capital*. Chicago: University of Chicago Press.

Harvey, David. 1989. *The Condition of Postmodernity: An Enquiry into the Origins of Cultural Change*. Cambridge, MA: Blackwell.

Hong, Seong-tae. 2011. *Togeon-gukgareul Gaehyeokhara: Gaebaljuuireul Neomeo Saengtaebokjigukgaro* [Reforming the Construction State: Beyond Developmentalism, Toward an Eco-Welfare State]. Paju, KR: Hanul Academy.

hooks, bell. 1986. "Sisterhood: Political Solidarity between Women." *Feminist Review* 23: 125–38.

Jung, Hyun Joo. 2005. "Scale in Contentious Politics: The Case of Greenbelt Controversy." *Journal of the Korean Urban Geographical Society* 8 (3): 121–37.

Katz, Cindi. 2001. "On the Grounds of Globalization: A Topography for Feminist Political Engagement." *Signs: Journal of Women in Culture and Society* 26 (4): 1213–34.

Kim, Jekook, and Tae-Kyung Kim. 2008. "Issues with Green Belt Reform in the Seoul Metropolitan Area." In *Urban Green Belts in the Twenty-First Century*, edited by Marco Amati, 37–57. London: Ashgate.

Lee, Hyunjoo. 1999. "Geurinbelteu Jeonmyeonhaejewa Siminundong-eui Daeeung" [The Wholesale Deregulation of Greenbelt and the Response of the Civil Society's Movements]. *Hwanggyonggwa Gonghae* [Environment and Pollution] 34.

Lee, Yong Sook, and Hae Ran Shin. 2012. "Negotiating the Polycentric City-region: Developmental State Politics of New Town Development in the Seoul Capital Region." *Urban Studies* 49 (6): 1333–55.

Lefebvre, Henri. (1970) 2003. *The Urban Revolution*. Minneapolis: University of Minnesota Press.

McCormack, Gavan. 1995. "Growth, Construction, and the Environment: Japan's Construction State." *Japanese Studies Bulletin* 15 (1): 26–35.

Mohanty, Chandra Talpade. 2003. "'Under Western Eyes' Revisited: Feminist Solidarity through Anticapitalist Struggles." In *Feminism without Borders: Decolonizing Theory, Practicing Solidarity*, edited by Chandra Talpade Mohanty, 221–51. Durham, NC: Duke University Press.

MOLTMA (Ministry of Land, Transport and Maritime Affairs). 2011. *Gaebaljehanjiyeok 40 Nyeon* [Development Restriction Areas, 1971–2011].

Oh, Seongkyu. 2000. "Teukjip: Han-guk Hwan-gyong-undong-ui Saeroun Mosaek 3 (National Trust Movement) – Simini Judohaneun Hwanggyeong, Munhwajasanui Sahwoejeok Gong-yuhwa Undong" [Special Issue: In Search of New Environmental Movement in Korea 3 (National Trust Movement) – People Initiated Movement to Make Environmental and Cultural Assets Social Commons]. *Environment and Life* 25: 40–9.

Orzeck, Reecia, and Laam Hae. Forthcoming. "Restructuring Legal Geography." *Progress in Human Geography*. https://doi.org/10.1177/0309132519848637

Palat, Ravi Arvind. 1999. "Fragmented Visions: Excavating the Future of Area Studies in a Post-American World." In *After the Disciplines: The Emergence of Cultural Studies*, edited by Michael Peters, 87–126. Westport, CT: Bergin and Garvey.

Park, Bae Gyoon. 2011. "Territorial Politics and the Rise of a Construction-Oriented State in South Korea." *Korean Social Sciences Review* 1 (1): 185–220.

Robinson, Jennifer. 2006. *Ordinary Cities: Between Modernity and Development*. London and New York: Routledge.

Roy, Ananya. 2009. "The 21st-Century Metropolis: New Geographies of Theory." *Regional Studies* 43 (6): 819–30.

Ruddick, Sue, Linda Peake, Gökbörü S Tanyildiz, and Darren Patrick. 2017. "Planetary Urbanization: An Urban Theory for Our Time?" *Environment and Planning D: Society and Space* 36 (3): 387–404.

Sayer, Andrew. 1991. "Behind the Locality Debate: Deconstructing Geography's Dualisms." *Environment and Planning A* 23: 283–308.

Shatkin, Gavin. 2007. "Global Cities of the South: Emerging Perspectives on Growth and Inequality." *Cities* 24 (1): 1–15.

Shin, Hyun Bang. 2009. "Property-based Redevelopment and Gentrification: The Case of Seoul, South Korea." *Geoforum* 40 (5): 906–17.

Sohn, Nakgu. 2008. *Budongsan Gyegeupsahoe* [The Real Estate Class Society]. Seoul: Humanitas.

Yang, Youngyeu. 1999. "Geurinbelteu Salligiundong Hwaksan … Simindanche Baegyeogot Hanmoksori" [The Expansion of the Movement to Save the Greenbelt … One Voice from about 100 Movement Organizations]. *Joong Ang Daily*, 7 August.

4 Marriage Migration as Spatio-Temporal Fix in Pohang's Post-Industrial Urban Development through Saemaul

HYESEON JEONG

"Asia as Method" for New Knowledge

Through an empirically grounded case study of a core location in Pohang, South Korea, this chapter demonstrates how "Asia as method" and Marxist theories can mutually expand each other. In applying David Harvey's spatio-temporal fix alongside Baik Yeong-seo's twofold-peripheral perspective to an analysis of Pohang's post-industrial urban development, I show that Asia as method does not dismiss Western theories in Asian studies but critically engages with them to create new knowledge about the world that does not marginalize Asia.

Theories are the building blocks from which our understanding of the world is constructed. Most theories, however, have been inspired by the experiences of the West and build a world modelled after it. The resulting geographical hierarchies in academia marginalize the non-West and dismiss area studies for their supposed particularism. It was expected that postcolonial studies would address the lived experiences of the non-West; however, given its obsession with the object of its critique (i.e., the West), postcolonial studies fails to transcend the academic geographical hierarchies between the West and the non-West (Chen 2010, 2). Area studies also falls short in this respect because it represents the knowledge produced by the West about the non-West more than that produced from within non-Western areas. It is with these issues in mind that Kuan-Hsing Chen asserts the need to critically deimperialize theory. One way to do so is to recognize the West as one cultural resource among many others, rather than as one of universal value, and to promote the production of knowledge that contributes to a relational understanding of culture. For scholars of Asian studies working in Asia, the alternative Chen suggests is "using the idea of Asia as an imaginary anchoring point, [through which] societies in Asia can become each other's points of reference, so that the understanding of the self may be transformed, and subjectivity rebuilt" (Chen 2010, 212). Imagining Asia as an anchoring point is certainly not a call to privilege Asia or to reinforce regional boundaries. It is a practical point of departure for Asian studies scholars to broaden their horizons of knowledge by mobilizing the diverse and yet shared histories and practices of their region. In particular, Chen suggests Asia as method as an attempt to overcome the legacies of colonialism, imperialism, and the Cold War through co-referencing among Asian societies as to how they deal with the specific forms of these legacies in their respective societies.

Building on the insights of Chen and other scholars engaged with the idea of Asia as method, Baik Yeong-seo promotes research from a twofold-peripheral perspective, which could contribute to creating "an autonomous space through which we can move beyond the trinity of the post-colonial, post–Cold War and post-hegemonic order, both theoretically and practically" (Baik 2013b, 145). The perspective of the twofold periphery, Baik argues, can be acquired from places marginalized by both global and national hierarchies. Some examples of such places that Baik uses include Dokdo, where territorial rivalry is taking place between South Korea and Japan; Okinawa, a Japanese island occupied by the US military; and Kinmen, the military front-line between Taiwan and China. These are places that are contested by geopolitical powers and are instrumentalized by their own national governments for their strategic interests. Throughout the twentieth century, the ideologies and practices of the core – including colonialism, imperialism, statism, militarism, as well as anticommunism – have largely dictated the fates of these peripheral places. Baik thus argues that these sorts of places are embedded in multiple layers of marginality, revealing the contradictions of the logic of the core and imbuing the possibility for resistance to its logic (Baik 2013a, 31). He labels these places "core locations" of research, which allow scholars to explore alternative ways of producing knowledge (Baik 2013b, 148). While Chen sheds light on Asia for the possibility of co-referencing in search of new politics, Baik focuses on East Asia for its geopolitical imperatives and potentials for critical regionalism. The interconnected history of the geopolitical and geo-economic contradictions in the core locations of East Asia, ironically, opens up the possibility of constructing an alternative community that overcomes ethnic nationalism and closed regionalism (Baik 2013b). Although Baik's examples of core locations are all found in geopolitical frontiers, he also argues that, in the search for new knowledge, a core location can be found in any place in the world where we can acquire a perspective of the twofold periphery (Baik 2013a).

Inspired by Baik's insight and at the same time critically examining the limitations of the notion of core location, I exercise the twofold-peripheral perspective in Pohang, South Korea to critically investigate the margins of this industrial city. Admittedly, Pohang has little geo-political or geo-economic stature and demonstrates no resistance to the order of colonialism, imperialism, or the Cold War. Quite to the contrary, Pohang – or, more accurately, urban areas of Pohang – was a beneficiary of the geopolitical and geo-economic order of the core when South Korea relied on the city's steel production for rapid industrialization during the 1970s and 1980s. After the industrial boom had swept the city, however, it was left with no further momentum for growth. Finding itself in the midst of fierce competition with other cities for global status, Pohang attempted to forge a post-industrial urban identity through a variety of programs. This chapter investigates one of these programs, Global Saemaul, which conflated Pohang's rural experience of an anti-poverty campaign with its urban experience of steel-oriented industrialization in order to create an international development aid program. Ironically, the program revealed and amplified the marginalities of Vietnamese women who had migrated to Pohang to marry when it made the women's natal families in Vietnam its targeted beneficiaries. After analysing the data collected through interviews and archival research conducted in 2016 and 2017, I argue that the ironies of Global Saemaul reveal not only the paternalistic and patrilocal logic of developmentalism that is embedded in Pohang's international development aid program, but also the regional, national, and global complexities of the socio-economic consequences of uneven development.

Marriage Migration as Spatio-Temporal Fix

Transnational migration has been predominantly understood as an issue of territoriality. Migrant workers, unless they are highly qualified, are often viewed as a threat that could potentially burden a destination country economically, politically, and socially. A competent government should have firm control over its borders and clear policies regulating migration. A proactive government would make an effort to address the push factors of migration in origin countries, as did many European governments with foreign aid in the 1990s. Recently, migrant remittances have grown substantially larger than foreign aid and more stably than private capital flows, revealing the high potential of migration for promoting development in origin countries (World Bank 2016). Migration facilitates not only financial transfers but also knowledge sharing and human capital exchange, creating new opportunities and possibilities. Migrants are now recognized as an important "resource" for promoting economic development (Nyberg-Sorensen, Van Hear, and Engberg-Pedersen 2002). The migration-development nexus thesis highlights migrants as transnational subjects who promote the "co-development" of both origin and destination countries (Faist 2008; Bailey 2010; Fauser 2014).

Hein de Haas (2010) cautiously points out that this economic optimism toward migration replaces the structural discourses of world-systems analysis. The migration-development nexus thesis abstracts transnational migration from broader transformations caused by globalization, thereby obscuring the relationship between the causes and effects of migration and neoliberal policies implemented in the name of development. By extension, the causes of migration (underdevelopment) are artificially separated from the effects of migration (development). In other words, the positive perspectives on migration taken by the migration-development nexus thesis are symptomatic of developmentalism. Although this literature sheds new light on the agency of migrants, it focuses narrowly on the forms of migration that are functional for economic growth. Parvati Raghuram (2009) brings our attention to the power of developmentalism and how the circulation of migrants keeps the idea in motion. Various paradigms of development have come and gone, but developmentalism has yet to be seriously challenged, especially by migrants themselves. Rather, migrants are expected to realize their potential and moral responsibility to overcome the limitations of foreign aid and the states of their origin countries in promoting development: "The mobile governable subject of migration-development ... is both required to move in order to strategise their human capital, but also to act morally for the collective good of a distant place/community" (Raghuram 2009, 110). For migrants, development is a matter of both agency and morality.

Arguably, one of the most gendered and ethnically laden forms of migration is marriage migration. An increasing number of Southeast Asian women are migrating to East Asia for marriage, and their remittances are an important source of income for their families back home. One of the popular narratives explaining women's decision to pursue transnational marriage migration is to improve their economic status and to help their natal families. Women make remittances not only because they feel pressure from their natal families to be filial daughters, but also for many other reasons, such as maintaining and strengthening family ties across borders and signalling their successful marriages (Thai 2008; Yeoh et al. 2013). Nonetheless, marriage migrants are largely absent from the migration-development nexus literature, because they are not considered to be workers and, by extension, agents of development. The androcentric economism of the migration-development nexus literature has highlighted the European experiences of labour migration and refugees and has overlooked the Asian experiences of transnational marriage and its exploitation of women's unofficial labour.

The feminization of intra-Asian marriage migration differs from the general feminization of migration, which is influenced by women's labour participation and the commodification of care work and emotional labour. Intra-Asian marriage migration is primarily a rural-to-rural migration process from Southeast Asia to East Asia through introduction services that were catalysed by the male marriage squeeze in East Asia as well as by patrilocal practices and women's hypergamy (Wang and Chang 2002; Belanger and Wang 2012; Liu, Brown, and Feldman 2014).[1] In a sense, the intra-Asian marriage migration of women is similar to the older Western phenomenon of mail-order brides: both are instances of transnational hypergamy mediated by introduction services. The two are different, however, in the sense that the former is promoted by East Asian states as a solution to their low fertility rates (Yang and Lu 2010, 17). Indeed, popular destination countries of marriage migration in East Asia are countries with the lowest total fertility rates in the world: Japan (with a fertility rate of 1.41%), South Korea (1.26%), Hong Kong (1.19%), and Taiwan (1.13%).[2] Transnational marriages in Japan have been on the rise since the 1970s, particularly through the marriage migration of Chinese and Filipina women. Taiwan witnessed the most rapid increase of transnational marriages in the world between the 1980s and the 2000s through the migration of mainly Indonesian, Vietnamese, and Chinese women. Hong Kong and South Korea followed in their footsteps soon thereafter.

Considering that this trend has been catalysed by state action, I argue that intra-Asian marriage migration represents a spatio-temporal fix aimed at social stability and reproduction. David Harvey's concept of spatio-temporal fix refers to solutions to capitalist crises through temporal deferral and geographical expansion (Harvey 2003, 115; 2006, 427). When profit rates fall, surplus capital is invested in long-term

1 "Marriage squeeze" refers to a disproportionate ratio between the number of males and females at the prime age of marriage; "hypergamy" is the practice of marrying into a group with a higher economic and social status.

2 The World Factbook, "Total Fertility Rate (2017 Estimates)," accessed 1 May 2018, https://www.cia.gov/library/publications/the-world-factbook/rankorder/2127rank.html

projects of geographical expansion and spatial reorganization to avoid devaluations. Bridges, roads, dams, and ports are popular examples of spatio-temporal fixes that provisionally solve crises of over-accumulation while providing new momentum for growth. In the age of global capital, the export and import of surplus capital, commodities, and labour power also work as spatio-temporal fixes that temporarily defer the crises beyond territorial boundaries (Jessop 2006, 162). The state would occasionally permit flows of migration in response to capital's need for cheap labour, but at other times would restrict them to assuage public fears of possible migrant-induced social problems (Scott 2013, 1092). The specific forms of spatio-temporal fixes through migration are influenced not only by geopolitical economy but also by migrants' intersectional factors such as gender, age, class, and ethnicity. As I demonstrate below through the case of Vietnamese women's marriage migration to South Korea, marriage migration and the subsequent mobilization of their care work can also act as a spatio-temporal fix for social problems in the destination country, such as population decline.

The South Korean state has promoted transnational marriage migration to address the problems of uneven development and population crisis. The country's urban-biased industrialization of the 1960s and 1970s resulted in substantial urban migration, which led to the ageing of rural communities as well as a female deficit in rural areas that was exacerbated by a traditional (prenatal) preferences for sons. Consequently, a chronic marriage squeeze occurred in rural areas. Rather than address the fundamental problems that rural communities faced, the state has encouraged rural Korean men to marry ethnic Korean women from China by lowering the territorial barriers to marriage migrant women and sponsoring the men's expenses associated with transnational marriage arrangements. This suggests that transnational marriage in South Korea is located more in the realm of governance than in the private sphere (H.M. Kim 2006). It is a spatio-temporal fix for the social consequences of urban-biased uneven development. At its peak in 2005, over 35 per cent of marriages registered in rural communities were transnational.[3]

The state did not consider the marriage squeeze to be a class issue or a rural problem, but rather a national population crisis (H. Lee 2012). The country's total fertility rate had been near the lowest in the

3 Statistics Korea, *In-gu donghyang josa* [Demographical Changes], 2005. Urban lower-class men are increasingly getting married to foreign women through introduction services, supporting the class-specific pattern of transnational marriage.

world since the late 1990s. Declining marriage rates were interpreted as a cause of the population crisis, and transnational marriage migration was promoted as a solution. Policies on marriage migration were reconceptualized as policies for "multicultural families" (*damunhwa gajok*). Marriage migrant women are colloquially referred to as "foreign daughters-in-law" (*oegugin myeoneuri*), implicitly highlighting their functionality within the marital family as care providers. While the government euphemistically calls them "multicultural women" (*damunhwa yeoseong*), its exclusive application of the idea of multiculturalism to marriage migrants, and not to other types of migrants in South Korea, suggests that the statist discourse of multiculturalism is not about social diversity but about marriage migrants' reproductive functionality (Oh 2007; Yoon 2008). (See also Eom's chapter in this volume for a discussion of how the Chinese residents in South Korea are marginalized by the statist discourse of multiculturalism.) In its present incarnation, Korean "multiculturalism" is a population policy because, in the name of promoting multiculturalism, the government controls the flow of transnational marriage migrants and Koreanizes transnational families (H.M. Kim 2014, 198).

The stop-gap fix of marriage migration contains within it many problems. Outcomes of state-sponsored transnational marriage migration have included a rapid increase in the number of commercial introduction service agencies. These agencies often infringe on women's human rights and limit their access to correct information about life in South Korea, as they prioritize the interest of their paying customers – that is, South Korean men (So 2009; Kwan and Kang 2016). Many marriage migrant women enter into marriage based on incorrect or false information about their spouses.

Hypergamy encourages transnational marriage migration, but transnational marriage migration does not automatically provide women with upward mobility. Rather, it often results in a deterioration of women's social status because these women's educational attainment, linguistic abilities, and racial profile are generally considered inferior to the average in destination countries, something Nicole Constable (2005) calls the paradox of global hypergamy. Marriage migrant women are often in a dependent position and have to negotiate remittances to their natal families with their husbands, though women with greater social capital are more likely to succeed in these negotiations (Belanger, Linh, and Duong 2011). The women are situated in a transnational terrain of patriarchy, attempting simultaneously to live up to the expectations placed on a wife to take care of the marital family and of a filial daughter to support the natal family (Yeoh et al. 2013, 446). Almost 10 per cent of clients needed help as a result of domestic or sexual violence, while 13.8 per cent consulted an agency about domestic disputes.[4] According to the 2015 National Survey of Multicultural Families, 70 per cent of the marriage migrant respondents considered their marital families to belong to a lower-middle class (33.3 per cent) or a low class (37.5 per cent), and 33.3 per cent were experiencing economic difficulties.[5] Many marriage migrants are burdened with the financial difficulties of both their natal and marital families and seek employment to remedy them. This often causes tension between transnational couples. Most husbands of marriage migrants are much older than their wives, and they fear that their wives might leave them once they become economically independent. In a survey conducted by the Korea Institute for Health and Social Affairs in 2008, 30.7 per cent of marriage migrants and 32.9 per cent of their Korean spouses identified economic difficulties as the primary cause of domestic disputes (Y. Kim 2008, 183). In 2016, transnational couples accounted for 7.7 per cent of the total marriages and 9.9 per cent of the total divorces.[6] According to the Korea Legal Aid Center for family relations, divorces among couples of transnational marriages are markedly rising, and the increase in transnational marriages and the economic difficulties associated with them are suspected to be the primary cause of this phenomenon (Korea Legal Aid Center for Family Relations 2017).[7]

Pohang's Uneven Development and Saemaul

Pohang is South Korea's major industrial city and home to one of the world's largest steel producers, POSCO. As the majority of the city's population is involved in businesses that cater to the steel producer, Pohang is arguably South Korea's most representative company town. In *Asia's Next Giant* (1989), Alice Amsden depicts POSCO as a primary

---

4 "Jangnyeon ijuyeoseong sangdam yocheong 10-myeong jung 1-myeong pongnyeok pihae" [One in Ten Migrant Women Experience Violence], *Seoul Shinmun*, 22 March 2018, http://seoul.co.kr/news/newsView.php?id=20180323009004&wlog_sub=svt_023

5 Ministry of Gender Equality and Family, *Jeon-guk damunhwagajok siltaejosa bunseok* [Analysis of the 2015 National Survey of Multicultural Families], 2016, http://www.mogef.go.kr/mp/pcd/mp_pcd_s001d.do?mid=plc503

6 Statistics Korea, *In-gu donghyang josa* [2016 Demographical Changes], 2017, http://kosis.kr/index/index.do

7 For further discussion on marriage migrant women, consult the rich literature on migration and gender in South Korea, including Freeman 2011; Jung 2012; M. Kim 2013; H.M. Kim 2014; D.Y. Kim 2017; H. Lee 2014; Piper and Lee 2016; and Choo 2016.

example of state-capital collaboration for late industrialization. The developmental state of the Park Chung-Hee regime (1961–79) envisioned that an integrated steel mill – that is, steel works that have all the functions necessary for producing, casting, and rolling both iron and steel – would provide a springboard for other industries and facilitate the country's economic growth. This is well captured in POSCO's mission statement that defined steel production as a symbol of national power (*cheolgang-eun gungnyeok*). In addition, Park's military government placed great importance on steel production for South Korea's security in light of the country's armistice with North Korea. For the first three decades, while the state owned the company, the government wholeheartedly supported the company's operations. It financed the inception and expansion of POSCO, subsidized its energy use, and granted it tax discounts. By reinvesting most of its profit in production facilities and personnel, POSCO rapidly increased its productivity and laid the foundation for the country's other heavy industries, such as shipbuilding, automotive, and machinery.

One aspect that Amsden and other developmental state theorists overlooked in their praise of POSCO's and South Korea's rapid growth was that this celebrated growth stemmed from the state's spatial selectivity.[8] During Japanese colonial rule (1910–45), investments in industrialization were concentrated in the Seoul metropolitan area and the Gyeongsang provinces across the Korea Strait from Japan. The Park regime's industrialization strategy continued investments in these already industrialized regions. At the end of the Park regime in 1979, 84.5 per cent of national manufacturing employment was concentrated in these regions (B.-G. Park 2008, 54). In addition, given that the Gyeongsang provinces were the birthplaces of President Park and many of his core staff members, many high-level positions in the state apparatus were filled by members of the Gyeongsang elite, who acted as regional ties to Park and his staff. Indeed, the Gyeongsang provinces have continued to be a stronghold of the Park-family regimes and the conservative party ever since. For example, when Park's daughter Park Geun-hye ran for president in 2012, she won 81 per cent and 63 per cent of the vote in North and South Gyeongsang Provinces, respectively. Pohang, North Gyeongsang (hereafter Gyeongbuk), was one of

8 The spatial selectivity of the developmental state created deep-rooted inequality and antagonism between regions in South Korea, as Oh's chapter in this volume demonstrates. For further discussion on South Korea's spatial strategies and uneven development, see Chung and Kirkby 2002; B.-G. Park 2008; and B.-G. Park and Gimm 2013.

the cities that benefited the most from the Park regime's investments in the 1960s and 1970s.

It is noteworthy that the investments in Pohang were made possible at the expense of providing compensation for the victims of colonial mobilization. The Park regime's pursuit of steel production was in opposition to the post-war world economic order envisioned by the United States. The US government considered the steel mill plan to be too ambitious for South Korea and favoured the country's focusing on labour-intensive, export-oriented light manufacturing (Rhyu 2003). The US opposition obstructed the Park regime's original plan to finance the construction of the steel mill through foreign aid. Instead, the regime decided to utilize Japanese colonial reparations for the creation of POSCO. Unlike the US government, the Japanese government welcomed the investment opportunity, as it opened up a new export market for Japanese machinery as well as the possibility of regionally expanding the operations of Japanese businesses (Ozawa 1979). The investments were made at the expense of the victims of colonial mobilization because the Park regime had removed the right of individuals to claim reparations when negotiating the terms of colonial reparations with Japan in 1965. Instead, the government collected US $800 million from Japan in grant and loans as comprehensive compensation. This was contrary to the government's earlier stance toward the victims of forced labour and military mobilization when it conducted a national survey of colonial victims in preparation for the negotiation of reparations with Japan. The regime's rationale for the government to receive comprehensive compensation on behalf of the victims stemmed from the belief that the colonial reparations should be used to develop the national economy and boost the nation's pride (Han 2014). The victims of colonial labour and military mobilization have been struggling to reclaim their right to individual reparations to this day.

The unevenness of postcolonial resource distribution could also be witnessed within Pohang itself. That city had not been large enough to sustain POSCO, so the company relied on urban migration. Besides, the company was reluctant to recruit native Pohang residents, as the nature of steel making requires people with education and experience. In the decade following the company's opening in 1968, the old harbour city's population tripled, from less than 70,000 to over 200,000 (S.O. Park 1992). POSCO employees constituted Pohang's new regional elite. The company's social investments were limited to the welfare of its employees. High-end apartment complexes were built and made available only to POSCO employees, who were also offered long-term, low-interest housing loans. Their children were schooled separately from those of other Pohang residents. While POSCO's profits made Pohang rich, it became a "divided city" between POSCO and the rest of Pohang (Jun 2011; Chang 2013).

Pohang prospered thanks to the steel industry in the 1970s and 1980s, but it encountered economic challenges in the 1990s. POSCO brought in no further investment once the country's second integrated steel mill was built elsewhere, following the government's attempt to address uneven development. The city's population grew older, and the economy started to slacken. Moreover, Pohang was merged with a neighbouring county, Yeong-il, which was predominantly rural and did not have any local economic specialties other than producing semi-dried herring. The merger was made in accordance with the government's administrative reconfiguration. In 1995, the government created so-called urban-rural integrated cities by merging financially disadvantaged rural counties and their neighbouring cities. Following the merger, Pohang's fiscal self-reliance ratio dropped below 40 per cent and has yet to recover.[9] To overcome the economic downturn, Pohang has pursued a form of post-industrial urban development. Investments were solicited to host research and development projects, as well as to create a technology park for innovative businesses and an economic free zone.

Another of the city's post-industrial urban development strategies was to develop heritage tourism by investing in one of the city's main cultural assets, Saemaul Undong. Saemaul Undong (literally translated as New Village Movement, hereafter abbreviated as Saemaul) is a rural development campaign that was spearheaded by the Park Chung-Hee regime from 1971 to 1979 to address the urban-rural development gap triggered by rapid industrialization. Aimed at modernizing rural infrastructure and increasing agricultural productivity, the campaign propagated the value of "self-help" to rural communities and encouraged them to volunteer their labour and provide resources to the community. In the first year of Saemaul, the small rural village of Munseong in Yeong-il county received a presidential award for the best practices of the campaign. The village had renovated old thatched houses, constructed new roads, built irrigation canals to increase rice yields, and raised chickens to generate community income to fund further projects. On 17 September 1971, President Park personally visited Munseong to award it with prize money and immediate electrification, declaring that

9 Gyeongbuk Provincial Government, *Jaejeong gongsi* [Financial Report], 2017, http://www.gb.go.kr/Main/open_contents/section/finace/page.do?mnu_uid=2683&LARGE_CODE=330&MEDIUM_CODE=20&SMALL_CODE=90mnu_order=3

the whole country should follow its example. This event has allowed Munseong to claim to be the birthplace of Saemaul.

In 2009, the Pohang government opened a Saemaul history museum in Munseong in honour of the award the village received in 1971. The museum was designed more to help Pohang become a global tourist destination than to preserve the history of Saemaul, as illustrated by the following statement made by Councillor Lee Sangbeom: "People from all over the country and beyond, particularly from China and Vietnam, are visiting Munseong to learn the Saemaul spirit. But they leave disappointed because there is no formal display of historical records ... Pohang needs the momentum to reignite Saemaul for the 21st century. By creating a museum of Saemaul in Munseong, we can make the people of Pohang proud, teach our history to the youth, attract tourists and increase regional income."[10] This statement conveniently conflates Munseong and Pohang and ignores the urban-rural disparities. The first half of the statement focuses on Munseong and the value of its history, but the second half turns its attention to Pohang and its economic development. In addition, the first half suggests that the so-called Saemaul spirit of self-help remains in Munseong, while the second half stresses the need to "reignite" it in Pohang. Understandably, Pohang as an industrial city had a different experience of Saemaul than that of rural villages such as Munseong. Cities participated in Saemaul in the later years of the Park regime, but their Saemaul was focused on disciplining factory workers and increasing productivity and had little to do with the ever-emphasized Saemaul spirit of self-help. In short, there was little ground for Pohang to construct an identity based on Saemaul, had it not been for its merger with Yeong-il county in 1995. Pohang required a new asset on which to capitalize to overcome stagnation and promote post-industrial development, and it appropriated Munseong's experience of Saemaul for this purpose.

The Pohang Saemaul Museum replaces the history of the divided city with that of an imaginary place where Saemaul developed a global city out of rurality. A visit to the museum begins on the first floor with a walk through the "Tunnel of History." The tunnel displays images of people suffering from hunger and poverty during the turbulent times of Japanese colonial rule (1910–45) and the Korean War (1950–53). At the end of the tunnel is a visual illustration of the so-called Saemaul spirit, suggesting that Saemaul put an end to the nation's sufferings. South Korea's economic growth was the result of rapid industrialisation, of

10 Pohang City Council, *Bonhoeui hoeuirok* [Minutes of the General Meeting] (125/2), 4 September 2006.

course, and cannot be single-handedly attributed to Saemaul. Whether Saemaul succeeded in making rural villages prosper after the 1970s is debatable, because urban-rural disparities continued to increase after Saemaul and many rural communities today suffer from escalating debt.

The highlight of the museum is the display of the historic visit of 17 September 1971. A miniature replica of the village illustrates the moment when President Park visited Munseong and listened to the villagers' presentation of their campaign outcomes. Glass display cases show the daily logs that the village leaders kept on their implementation of Saemaul projects along with the award certificate they received from President Park. On the wall display rest pictures from the presidential visit along with the quotation "Make every village in the country a new village after the model of Munseong." The quotation is displayed several times throughout the museum in various forms.

Oddly enough, the museum does not explain whether or not the village continued practising Saemaul afterward, or how the village is situated vis-à-vis other rural villages in the country today. Munseong is represented as one of South Korea's poorest villages in the 1960s and the leading model of Saemaul in 1971 – no later details are provided. The story of the village ends on that historic day in 1971, concealing the fact that many villagers later left Munseong and its population decreased from 409 in 1970 to 220 in 2015.[11] For Pohang's new urban identity production, it did not matter what Munseong had become after Saemaul. The rest of the museum focuses only on the present and future of Pohang with its high-tech industries and state-of-the-art urban infrastructure. Pohang's industrialization and economic growth preceded the city's merger with Yeong-il county and cannot be attributed to Munseong's best practices of Saemaul. And yet, the museum presents a revised chronology that begins with Munseong's Saemaul project and ends with Pohang's global desires, misleadingly suggesting that Pohang had grown from a rural village to an aspiring global city through Saemaul. The revised chronology implies an erroneous causality between the Saemaul spirit and South Korea's economic development and another between Munseong's history of Saemaul and Pohang's prosperous future. In this narrative, Saemaul loses its meaning and becomes an empty signifier for Pohang's global urban identity.

Pohang's strategy of capitalizing on rural experiences for urban development is well captured in its lawsuit against Cheongdo, another

---

11 Pohang Municipal Government, *Tonggyeyeonbo* [57th Statistical Yearbook], 2017.

county in Gyeongbuk about 100 kilometres away from Pohang. Cheongdo's rural development projects and self-help spirit allegedly inspired President Park to design the Saemaul campaign when he was passing through the county in 1969. In the same manner as Pohang, Cheongdo commemorated the presidential visit by building a Saemaul history museum and developing Saemaul education tour programs. As Cheongdo also claimed to be the birthplace of Saemaul, Pohang sued Cheongdo for defamation in 2009. During the trial, to claim Saemaul as their identity, both parties eagerly presented evidence of Park Chung-Hee's visit to their respective villages and argued over the significance of the visits, instead of showing how actively they have practised Saemaul and developed their rural areas. This incident suggests that the value of being the birthplace of Saemaul for both parties stemmed from its potential for developing a tourist destination. The two museums were built more to anchor the national history of Saemaul in Pohang and Cheongdo, respectively, than to disseminate the Saemaul spirit globally. Pohang's Global Saemaul brought the past of rural areas to light, but it turned a blind eye to their present – that is, to what urban-centred development had bequeathed to them.

The Saemaul fever exemplified by Pohang and Cheongdo needs to be contextualized within the trend of globalizing Saemaul in the aftermath of the 1997 Asian financial crisis (Jeong 2017). Originally a state-led rural development campaign, Saemaul lost momentum at the end of the Park Chung-Hee regime partly because the campaign was involved in the corruption scandals of the succeeding regime, and partly because democratization put an end to the state apparatuses that had mobilized people's labour and resources for the campaign. When the Asian financial crisis placed the country in jeopardy through the disgrace of bankruptcy, people dusted off Saemaul in their memory and started community service provision in the name of Saemaul. For people who had participated in, or were educated about, Saemaul in the 1970s, community service provision was not only a solution to social problems but a way of contributing to national development (Jeong 2017). Some of the popular activities of the revived Saemaul include kimchi-making for disadvantaged households, giving baths to the elderly, and environmental beautification. Today, Saemaul is a community service movement led by the Korea Saemaul Undong Center, which has approximately two million due-paying members across the country. The revived Saemaul also expanded the geographical scope of its mission toward developing countries. It claims that the Saemaul spirit of self-help contributes to making developing countries self-reliant and independent of international aid. The resulting Saemaul projects abroad are a mixture of material aid and volunteer service provision that all international charity organizations adopt today, with rural development activities that the original Saemaul promoted in the 1970s. Saemaul volunteers visit rural areas in developing countries to teach new agricultural technologies and the Saemaul spirit of self-help, but they never leave a place without making donations of food, medicine, or electronic appliances. Saemaul's transformation from a self-help rural development campaign to a community service movement and the partial overlap of the two in the Saemaul projects abroad produce an ironically convergent discourse in South Korea. It is a discourse of national development that highlights the country's successful development and asserts the usefulness of its experience for promoting international development (Jeong 2017). The discourse of national development masks Saemaul's contradictory promotion of the ideal of self-help and the practice of service provision, as well as its paternalistic gaze on developing countries.

The Gyeongbuk provincial government spearheaded the trend of globalizing Saemaul. As explained earlier, the province had been a political stronghold of Park Chung-Hee and the conservative party more generally. By embracing Saemaul, one of the legacies of Park, politicians could appeal to conservative voters and benefit from the grass-root networks of Saemaul during elections. For example, Governors Lee Eui-geun (1995–2006) and Kim Kwan-yong (2006–18) enthusiastically promoted Saemaul, which could have contributed to their each being elected three consecutive times to the governorship. Governor Kim was particularly interested in globalizing Saemaul to replace its old image with a more updated one for the global era. One of the popular strategies that local governments implemented after the restoration of local autonomy was to build international networks and to "globalize" their cities and local businesses (J.-S. Lee and Woo 2010). Under the leadership of Governor Kim, the Gyeongbuk government thematized its international networks with Saemaul and provided developing countries with aid to disseminate its ideas. Calling himself "Mr. Saemaul," Governor Kim established the Global Saemaul Foundation to finance his government's Saemaul aid, and he travelled to rural villages in Asia and Africa to demonstrate how to practise Saemaul. Today, Gyeongbuk is by far the largest subnational donor of aid in South Korea (Cho, Park, and Jung 2015, 274). Every year, the Global Saemaul Foundation sends off hundreds of volunteers to developing countries and invites hundreds of trainees from target countries. Most of the trainees visit Pohang to learn about the history of Saemaul at the Pohang Saemaul Museum and tour one of South Korea's largest industrial facilities at POSCO. The national Saemaul wave since the late

1990s and the Gyeongbuk government's global Saemaul initiative provided a perfect opportunity for Pohang to utilize Munseong's historic achievement to claim Saemaul as the city's identity. Pledging to disseminate the Saemaul spirit to countries with difficulties to help eradicate poverty and hunger in the world, the city's mayor Park Seung-ho (2006–14) launched an international development aid program called Pohang Global Saemaul in 2011.[12]

Pohang's Saemaul and Marriage Migration

The encounter between Saemaul and marriage migrants in Pohang came at an unexpected time. When the Pohang Municipal Government (PMG) launched Pohang Global Saemaul, the program aimed at replicating Saemaul as a rural development campaign in developing countries such as Madagascar by transferring agricultural technologies as well as the so-called Saemaul spirit of self-help. Political conflict between the Gyeongbuk governor and Pohang mayor, however, forced the PMG to prematurely terminate the program in 2015. The program's unused budget was entrusted to the Pohang Saemaul Association (PSA), the regional chapter of Saemaul in Pohang that had been assisting the PMG with its Madagascan project. Searching for a way to utilize the remaining budget, the PSA learned that other Saemaul chapters in Gyeongbuk had projects that targeted marriage migrants or their natal families in their origin countries. Vietnamese marriage migrants make up to 39.2 per cent of Gyeongbuk's multicultural families, while, nationwide, they constitute only 21 per cent of multicultural families.[13] These numbers contributed to the PSA decision to turn its eye toward Vietnamese marriage migrant women.

As discussed earlier, many transnational families struggle with financial difficulties. In Pohang, as anywhere else, financial issues often lead to the dissolution of or violence in transnational families. In 2007, for example, a marriage migrant woman in Pohang was choked to death by her husband who had recently lost his job. As the woman became the sole breadwinner, the husband became anxious that she would

12 Pohang Municipal Government, *Geullobeol Pohang Bijeon 2020* [Global Pohang Vision 2020], 2010.

13 "Dayanghan gyeolhon iminjadeul saneun gyeongbukdoneun riteul woldeu" [Gyeongbuk Is a 'Little World' of Multicultural Families], *Kukmin Daily*, 22 January 2013, http://news.kmib.co.kr/article/view.asp?arcid=0006828596&code=11131418&sid1=all

leave him. Ultimately, he killed his wife and attempted suicide.[14] In response to these local problems, the PSA decided to use the PMG's Global Saemaul budget and help attenuate the financial burdens of transnational families. The PSA reasoned that transnational marriages often fail because marriage migrants are preoccupied with supporting their natal families in their countries of origin. For a transnational couple to lead a happy married life, it was believed, the migrant woman should be freed from concerns about her natal family's economic situation and feel at home in Pohang. Promoting the "happiness" (*haengbok*) of marriage migrants' natal families in Vietnam was considered the same as promoting the happiness of the migrants and their families-in-law in Pohang.

The central activity of the PSA's Vietnam project is housing provision for migrant women's natal families (*chinjeongjib jieojugi*). Constructing new houses or renovating old houses for migrant women's natal families is not unique to Saemaul – it has been performed by many organizations, including the Korean Red Cross and the Korea Land and Housing Corporation, as a form of international charity. For example, when Typhoon Haiyan hit the Philippines in 2013, destroying thousands of houses and killing just as many people, donations were collected across Korea to help rebuild the houses of the natal families of Filipina marriage migrant women. What distinguishes the Saemaul housing project from other housing projects for the natal families of marriage migrants, though, is the connection local Saemaul chapters have with the beneficiary marriage migrants and the relationship between the project and other Saemaul activities. House renovation was a popular activity during the original Saemaul campaign in the 1970s, when the government encouraged replacing straw-thatched roofs with slate roofs. Villagers cooperated to renovate homes one by one until the entire village had new roofs. Today's Saemaul as a community service movement continues to do house renovation for low-income families. Saemaul members volunteer their time and skills to change old floorings, fix leaking roofs, and insulate the thin walls of houses occupied by lone elderly people or those with disabilities. The PSA's Vietnam project is an international extension of such house renovation services for domestic low-income families. The project also includes other activities

14 Dong-u Shin "Joseonjok yeoseong u-uljeung nampyeone mokjollyeo salhae" [Ethnic Korean Chinese Woman Killed by Depressed Husband], *Gyeongbuk Maeil*, 27 December 2007, http://www.kbmaeil.com/news/articleView.html?idxno=37999

such as donating educational equipment to a local school in Vietnam, giving lectures about the Saemaul spirit, and building a chicken farm for income generation. Still, the major activity of the project is housing provision. When launching the project in 2016, the PSA hired a local contractor to be in charge of the half-year-long construction while its members made frequent visits to the project site to evaluate the progress and volunteer their labour and skills for the construction work.

Advertising for the project is assisted by the PMG's Multicultural Family Support Center (hereafter the Multicultural Center). The Multicultural Center is a government agency that provides marriage migrants with Korean language classes, cultural orientations, marriage counselling, parenting lessons, as well as social networking opportunities. At the request of the PSA, the Multicultural Center advertises the housing project on its walls, website, and online networking service (Naver Band) and then collects applications. In addition, it helps the PSA shortlist the candidates.

The PSA interviews each applicant, when possible, together with her husband. Each application is evaluated based on the economic conditions of the natal family and marital family, as well as the marriage life of the couple. In 2016, the PSA shortlisted four candidates whose natal families had significantly poor living conditions, even by Vietnamese standards. The candidates' marital families in Pohang were deemed to have either a middle or middle-low standard of living. After a field visit in Vietnam, the PSA selected two of the four candidates from the same district in Dong Thap province and two more households in the same district to fill the remaining two spaces. Selecting four households from the same district was intended to minimize transportation costs and, at the same time, maximize the visibility of the project. Two of the four houses for renovation were too old and unstable to renovate, so they had to be rebuilt. Besides, the families had specific preferences for what they wanted in their new houses. With the families voluntarily covering the extra costs, the two houses were rebuilt from scratch, and one of the other two was expanded.

The PSA made four trips to Vietnam between August and November 2016. The first trip was a preliminary field survey conducted by two PSA executive members and one PMG official. They collected information on local construction costs and the logistics of operating between Pohang and the district in Vietnam and conducted a needs assessment of a local primary school that the two marriage migrants had attended while growing up. The second trip was made by one PSA executive member to make the final arrangements for the project and to sign contracts with the local construction contractors in the district. The third and fourth trips focused on participating in the construction process and building a chicken farm. Five male volunteers stayed in the district for eleven days to assist the construction contractors. The final trip included five male volunteers, four female volunteers, and the two marriage migrants. All travel costs were equally shared by the PSA and the participants. One of the volunteers ran a construction business in Pohang, and participated in both trips to offer his expertise.

The two marriage migrant women participated throughout the construction process by facilitating communication between the PSA and the families in Vietnam. Given their fluency in Vietnamese and Korean and familiarity with both cultures, the women played the role of local guide and interpreter during the PSA's fourth trip to Vietnam. The PSA also utilized a popular social network service (Kakaotalk) to communicate with the women's natal families in Vietnam. The families regularly took photos of the construction sites and transmitted them to the PSA via smartphones. The extensive involvement of the marriage migrants and their natal family members ensured that the project reflected the needs and desires of the families in Vietnam. It also helped the PSA develop a sense of community with the marriage migrants and their natal families. Compared to the conventional practices of foreign aid, including the South Korean government's Saemaul projects abroad, which select project sites and activities based on the donor's political and economic interests, the PSA's housing project in Vietnam is recipient-oriented and has the potential of building a lasting relationship of care between the donor and the recipient.

Nevertheless, the donor-oriented and paternalistic nature of the project is hard to overlook. Despite the self-claimed Saemaul spirit of self-help, the PSA's housing project offers few opportunities for the Vietnamese recipients to contemplate their own definition of development; rather, it replicates the practice of temporary relief associated with conventional foreign aid. In addition, the project reveals the patrilocal logic of marriage migration, since it explicitly aims at promoting Pohang's own development, by facilitating the settlement of marriage migrants, instead of that of Vietnam. The formal objective of the project is "to share the burden of marriage migrants to support their natal families and facilitate their resettlement in Pohang," but the choice of housing project implies much more. Renovating or rebuilding houses is a popular way of investing or displaying the wealth accumulated with migrant remittances in Southeast Asia (McKay 2005; Faier 2013; Peluso and Purwanto 2018). Many marriage migrants have the desire to perform their filial duty by "giving their natal parents a nice house," as labour migrants do with their remittances. The PSA chose a program of housing provision not only because the association had the skills and experience to pursue it, but because the activity could visually fulfil the filial duty of the marriage migrant. A renovated or rebuilt house symbolizes the presumably improved standard of living that the daughter might enjoy overseas, irrespective of the actual living conditions of her marital family. It also publicly showcases the benefit of having a daughter living in a developed country such as South Korea. Knowing that her natal family lives in the best house in her hometown, the daughter might be able to free herself from concerns about her natal family in Vietnam and better assimilate into life in Pohang.

Conclusion: Multiple Marginalities of Marriage Migrant Women in Pohang

Chen's Asia as method is a call to overcome the problems of Western-oriented knowledge production through co-referencing within Asian studies in Asia. It is neither a dismissal of Western theories nor an accord on Asia's superiority, as claimed by some Asian societies and governments. Rather, Asia as method is a constant reminder that Asian scholarship should work on revealing and remedying the violent legacies of colonialism, imperialism, and the Cold War. This chapter has shown how Asia as method and Marxist theories can mutually expand each other by employing Baik Yeong-seo's twofold-peripheral perspective. Baik has contributed to the discussion of Asia as method by bringing to our attention the places that are located at the junction of multiple layers of marginalization and that can furnish alternative sources of knowledge production and politics. Baik's methodology underpinned this research, which focuses on Vietnamese transnational marriage migrant women in Pohang who are situated at the transnational intersection of patriarchy and developmentalism. Facing the end of steel-induced development, Pohang tried to build a new urban identity based on the rural areas' history of Saemaul, regardless of the realities of these areas. Pohang's rural areas had been excluded from the city's industrial growth and experienced both a population decline and a marriage squeeze. The state presented transnational marriage migration as a solution to both.

This chapter has argued that such state-sponsored transnational marriage migration is a spatio-temporal fix for socio-economic problems of post-industrial stability and reproduction, which proactively utilizes territorial boundaries. The case of Pohang allows us to investigate not only the stop-gap way in which capital invests in built environments but also a variety of scale-jumping programs that attempt to provisionally remedy the socio-economic consequences of uneven development. More importantly, it enabled us to contemplate the gendered aspects of spatio-temporal fixes. As urban-rural disparities caused a chronic female deficit in rural areas and the nation's total fertility rate rapidly declined, the South Korean state selectively loosened territorial boundaries for marriage migrant women and provided services for their successful settlement and family raising. Migrants' labour and their moral responsibility to families and compatriots are mobilized and spatially rearranged to address the needs and concerns of both the destination and origin countries and delay crises of uneven development. Particularly, marriage migrant women are subject to the compound moral expectations of taking care of their marital families and sending remittances to their natal families. Pohang's Global Saemaul is an instance of international development aid being used to help marriage migrants meet these expectations while creating a new development momentum for the donor. The Saemaul housing project in Vietnam illustrates how international development aid is implicated in amplifying the marginalities marriage migrant women experience while also challenging them. Its goal of remedying the economic difficulties of the natal families of marriage migrant women was based on the fear they might interfere with Pohang's development. The uneven development that triggers transnational migration, however, is a structural one that cannot be easily ameliorated by cursory attempts such as building houses. The Saemaul housing project is yet another spatio-temporal fix for the challenges posed by the spatio-temporal fix of transnational marriage migration in Pohang.

Baik's twofold-peripheral perspective sheds light on the intersection of multiple marginalities at which Vietnamese marriage migrants in Pohang are situated. At the same time, the case of Pohang challenges Baik's focus on East Asia in his core location discussion. Baik's East Asia thesis is concerned with the historically entangled relations among China, Japan, and South Korea and the unending conflicts and misunderstandings that arise from those relations. Hence, his thesis emphasizes the potential of peace-building scholarship in the region. The imperial (or sub-imperial) projects of economic and military expansion of the three countries, however, increasingly challenge the validity of East Asia for practising Baik's twofold-peripheral perspective. While the East Asia thesis does not intend to promote regionalism, it inadvertently privileges what takes place within East Asia over events in other regions or between other regions and East Asia. Nevertheless, as illustrated in the case of Pohang and Vietnam, the twofold-peripheral perspective is effective in uncovering the new sorts of contradictions that globalization has produced beyond the familiar regional boundaries.

REFERENCES

Amsden, Alice Hoffenberg. 1989. *Asia's Next Giant: South Korea and Late Industrialization*. Oxford: Oxford University Press.

Baik, Young-Seo. 2013a. *Haeksimhyeonjang-eseo Dong-asiareul Dasi Mutda: Gongsaengsahoereul Wihan Silcheon-gwaje* [Rethinking East Asia in Core Locations: A Task for a Co-Habitation]. Paju, KR: Changbi.

Baik, Young-Seo. 2013b. "An Interconnected East Asia and the Korean Peninsula as a Problematic: 20 Years of Discourse and Solidarity Movements." *Concepts and Contexts in East Asia* 2: 133–66.

Bailey, Adrian J. 2010. "Population Geographies, Gender, and the Migration-Development Nexus." *Progress in Human Geography* 34 (3): 375–86.

Belanger, Daniele, Tran Giang Linh, and Le Bach Duong. 2011. "Marriage Migrants as Emigrants: Remittances of Marriage Migrant Women from Vietnam to Their Natal Families." *Asian Population Studies* 7 (2): 89–105.

Belanger, Daniele, and Hong-zen Wang. 2012. "Transnationalism from Below: Evidence from Vietnam-Taiwan Cross-Border Marriages." *Asian and Pacific Migration Journal* 21 (3): 291–316.

Chang, Sehoon. 2013. "Pohang: Gieobdosiui Sahoesaengtaehak" [Pohang: Socioecology of an Industrial City]. In *Gukgawa Jiyeok* [National State and Regions], edited by Bae-Gyoon Park and Dong-Wan Gimm, 198–227. Seoul: Alto.

Chen, Kuan-Hsing. 2010. *Asia as Method: Toward Deimperialization*. Durham, NC: Duke University Press.

Cho, Hyun Joo, Geon Woo Park, and Heon Joo Jung. 2015. "Han-guk Jibangjeongbuui ODA" [ODA Offered by South Korean Regional Governments]. *Jibanghaengjeongyeon-gu* [Regional Administration Research] 29 (1): 261–89.

Choo, Hae Yeon. 2016. *Decentering Citizenship: Gender, Labor, and Migrant Rights in South Korea*. Stanford, CA: Stanford University Press.

Chung, Jae-Yong, and Richard Kirkby. 2002. *The Political Economy of Development and Environment in Korea*. London: Routledge.

Constable, Nicole, ed. 2005. *Cross-Border Marriages: Gender and Mobility in Transnational Asia*. Philadelphia: University of Pennsylvania Press.

de Haas, Hein. 2010. "Migration and Development: A Theoretical Perspective." *International Migration Review* 44 (1): 227–64.

Faier, Lieba. 2013. "Affective Investments in the Manila Region: Filipina Migrants in Rural Japan and Transnational Urban Development in the Philippines." *Transactions of the Institute of British Geographers* 38 (3): 376–90.

Faist, Thomas. 2008. "Migrants as Transnational Development Agents: An Inquiry into the Newest Round of the Migration-Development Nexus." *Population, Space and Place* 14 (1): 21–42.

Fauser, Margit. 2014. "Co-Development as Transnational Governance: An Analysis of the Engagement of Local Authorities and Migrant Organisations in Madrid." *Journal of Ethnic and Migration Studies* 40 (7): 1060–78.

Freeman, Caren. 2011. *Making and Faking Kinship: Marriage and Labor Migration between China and South Korea*. Ithaca, NY: Cornell University Press.

Han, Hye-in. 2014. "Hanil Cheonggugwon Hyeobjeong Chegyeol Jeonhu Gangjedongwon Pihaeui Beomwiwa Bosangnolli Byeonhwa" [Changes in Estimating the Extent of Forced Mobilization and Legitimizing Reparations]. *Sahagyeon-gu* [Historical Research] 113: 237–83.

Harvey, David. 2003. *The New Imperialism*. London: Oxford University Press.

Harvey, David. 2006. *The Limits to Capital (1982)*. London: Verso.

Jeong, Hyeseon. 2017. "Globalizing a Rural Past: The Conjunction of International Development Aid and South Korea's Dictatorial Legacy." *Geoforum* 86: 160–68.

Jessop, Bob. 2006. "Spatial Fixes, Temporal Fixes and Spatio-Temporal Fixes." In *David Harvey: A Critical Reader*, edited by Noel Castree and Derek Gregory, 142–66. Malden, MA: Blackwell.

Jun, Sang-In. 2011. "Oesaengjeok Gieobdosieseo Hyeomnyeokjeok Gieobdosiro" [From an Exogenous City to a Cooperative City]. *Han-gukjiyeokgaebal* [Korean Regional Development] 23 (2): 1–18.

Jung, Hyunjoo. 2012. "Constructing Scales and Renegotiating Identities: Women Marriage Migrants in South Korea." *Asian and Pacific Migration Journal* 21 (2): 193–215.

Kim, Daisy Y. 2017. "Resisting Migrant Precarity: A Critique of Human Rights Advocacy for Marriage Migrants in South Korea." *Critical Asian Studies* 49 (1): 1–17.

Kim, Hyun Mee. 2006. "Gukjegyeolhonui Jeonjigujeok Jendeo Jeongchihak" [Gender Politics of Transnational Marriage]. *Gyeongjewasahoe* [Economy and Society] 70: 10–37.

Kim, Hyun Mee. 2014. *Urineun Modu Jibeul Tteonanda* [We All Leave Home]. Seoul: Dolbegae.

Kim, Minjeong. 2013. "Weaving Women's Agency into Representations of Marriage Migrants: Narrative Strategies with Reflective Practice." *Asian Journal of Women's Studies* 19 (3): 7–41.

Kim, Yugyeong. 2008. *Damunhwasidaereul Daebihan Bokjijeongchaekbang-an Yeon-gu* [Welfare Policies for a Multicultural Society]. Seoul: Korea Institute for Health and Social Affairs.

Korea Legal Aid Center for Family Relations. 2017. "Damunhwagajeong Ihonsangdam Tonggye" [Statistics on Interracial Divorce]. 25 April.

Kwan, Han-Woon, and Byeongro Kang. 2016. "Asia Damunhwa Yeoseong-ui Gukjegyeolhon Junggaegwajeong Munjewa In-gwonboho" [Asian Women's

Transnational Marriage and Human Rights]. *Ataeyeon-gu* [Asian Pacific Research] 23 (1): 31–60.

Lee, Hyunok. 2012. "Political Economy of Cross-Border Marriage: Economic Development and Social Reproduction in Korea." *Feminist Economics* 18 (2): 177–200.

Lee, Hyunok. 2014. "Trafficking in Women? Or Multicultural Family? The Contextual Difference of Commodification of Intimacy." *Gender Place and Culture* 21 (10): 1249–66.

Lee, Jeong-Seok, and Yang-Ho Woo. 2010. "Jibangjeongbu Gukjegyolyuui Yeonghyang-yoin" [International Exchange among Regional Governments]. *Jibanghaengjeongyeon-gu* [Regional Administration Research] 24 (4): 393–421.

Liu, Lige, Xiaoyi Jin, Melissa J. Brown, and Marcus W. Feldman. 2014. "Male Marriage Squeeze and Inter-Provincial Marriage in Central China: Evidence from Anhui." *Journal of Contemporary China* 23 (86): 351–71.

McKay, Deirdre. 2005. "Reading Remittance Landscapes: Female Migration and Agricultural Transition in the Philippines." *Geografisk Tidsskrift / Danish Journal of Geography* 105 (1): 89–99.

Nyberg-Sorensen, Ninna, Nicholas Van Hear, and Poul Engberg-Pedersen. 2002. "The Migration-Development Nexus: Evidence and Policy Options State-of-the-art Overview." *International Migration* 40 (5): 3–47.

Oh, Kyung Seok, ed. 2007. *Hangugeseoui Damunhwajuui* [South Korea's Multiculturalism]. Seoul: Hanul.

Ozawa, Terutomo. 1979. *Multinationalism, Japanese Style: The Political Economy of Outward Dependency*. Princeton, NJ: Princeton University Press.

Park, Bae-Gyoon. 2008. "Uneven Development, Inter-Scalar Tensions, and the Politics of Decentralization in South Korea." *International Journal of Urban and Regional Research* 32 (1): 40–59.

Park, Bae-Gyoon, and Dong-Wan Gimm, eds. 2013. *Gukgawa Jiyeok* [National State and Regions]. Seoul: Alto.

Park, Sam Ock. 1992. "Jecheolsaneobdosi Pohangui Baljeon-gwa Jeongchaekgwaje" [Pohang's Development and Its Future]. *Dosimunje* [Urban Question] 27: 45–57.

Peluso, Nancy Lee, and Agus Budi Purwanto. 2018. "The Remittance Forest: Turning Mobile Labor into Agrarian Capital." *Singapore Journal of Tropical Geography* 39 (1): 6–36.

Piper, Nicola, and Sohoon Lee. 2016. "Marriage Migration, Migrant Precarity, and Social Reproduction in Asia: An Overview." *Critical Asian Studies* 48 (4): 473–93.

Raghuram, Parvati. 2009. "Which Migration, What Development? Unsettling the Edifice of Migration and Development." *Population, Space and Place* 15 (2): 103–17.

Rhyu, Sang-Young. 2003. "Between International Constraints and Domestic Policy Choice: Korea's Economic Development in the 1960s." *Korean Political Science Review* 37 (4): 139–66.

Scott, Sam. 2013. "Labour, Migration and the Spatial Fix: Evidence from the UK Food Industry." *Antipode* 45 (5): 1090–109.

So, Rami. 2009. "Gyeolhon Ijuyeoseong-ui In-gwon Siltaewa Han-guk Beobjedo Hyeonhwang-e Daehan Geomto" [Marriage Migrant Women's Human Rights and South Korea's Legal System]. *Beobhangnonchong* [Law Journal] 16 (2): 1–32.

Thai, Hung Cam. 2008. *For Better or for Worse: Vietnamese International Marriages in the New Global Economy*. New Brunswick, NJ: Rutgers University Press.

Wang, H.Z., and S.M. Chang. 2002. "The Commodification of International Marriages: Cross-Border Marriage Business in Taiwan and Vietnam." *International Migration* 40 (6): 93–116.

World Bank. 2016. *Migration and Remittances Factbook*. Washington, DC: World Bank.

Yang, Wen-Shan, and Melody Chia-Wen Lu, eds. 2010. *Asian Cross-Border Marriage Migration: Demographic Patterns and Social Issues*. Amsterdam: Amsterdam University Press.

Yeoh, Brenda S.A., Chee Heng Leng, Vu Thi Kieu Dung, and Cheng Yi'en. 2013. "Between Two Families: The Social Meaning of Remittances for Vietnamese Marriage Migrants in Singapore." *Global Networks* 13 (4): 441–58.

Yoon, In-Jin. 2008. "Han-gukjeok Damunhwajuuiui Jeon-gaewa Teukseong" [Characteristics of South Korean Multiculturalism]. *Han-guksahoehak* [Korean Sociology] 42 (2): 72–103.

5 "Locations of Reflexivity": South Korean Community Activism and Its Affective Promise for "Solidarity"

MUN YOUNG CHO

This chapter situates anti-poverty community activism across different regions in Asia as a core location. Focusing on the discursive formation of "location of reflexivity," it critically engages in similar ideas discussed by scholars of "Asia as method." South Korean community activism has always been nourished through global dialogues and translations. As an associate director of the Institute on Urbanization at Yonsei University in 1968–70 and a training director of the Philippine Ecumenical Council for Community Organization in 1970–72, American pastor Herbert D. White helped early activists in South Korea and the Philippines learn the community organizing (CO) methodology of Saul D. Alinsky (1989).[1] Alinsky's principle that social change is impossible without the empowerment of the poor and their collective action greatly affected community activism in Asian countries. Organized by Asian bishops in 1971, the Asian Committees for People's Organization (ACPO) helped build institutions for training and managing community organizers in the Philippines, Hong Kong, Indonesia, Thailand, and India. In response to the forced demolition of shantytowns, Asian activists, including South Koreans, established the Asian Coalition for Housing Rights (ACHR) in 1988. They also built Leaders and Organizers of Community Organization in Asia (LOCOA) in 1993, as a successor of ACPO, in order to "introduce an extensive network of … [community organizations] and facilitate the exchange of CO tactics

This work was supported by the National Research Foundation of Korea (NRF) Grant funded by the Korean Government (MEST –NRF-2018S1A6A3A01081183).

1 The work of Saul D. Alinsky and Paulo Freire contributed to developing CO as the main method for grass-roots activism in Asian countries in the 1960–70s; see Inamoto (2011, 15–16). Yet, the method has revealed historical and regional differences in the form that activism takes, as it intersects with the changing praxis of political economy.

and experiences between activists in Asian countries."[2] Based on their long-term CO experiences, South Korean activists built the Korean Community Organization Network (KONET) as a centre for training in CO methods and continued to seek solidarity with LOCOA as members of KONET (KONET 2010, 14; Inamoto 2011, 47–50).[3]

Over time, though, many South Korean activists have found themselves in a dilemma as they pursue solidarity across Asian countries. Solidarity was made possible by common histories of violent eviction and deportation, which the urban poor in most Asian countries experienced in light of the rapid modernization, economic development, and political turbulence of the twentieth century. In a speech in 1950, Jawaharlal Nehru, the first prime minister of India, described such circumstances as "the torment in the soul of Asia": "Compared with other regions of the world," he said, "Asia was in the midst of the most drastic changes, yet it could not change slowly; the drastic changes were accompanied by danger but Asians had no choice, and this was the biggest torment for Asians" (quoted in Sun 2013, 221). However, this "torment" has increasingly become a narrative of the *past* in South Korea, where democratization movements in the 1980s eventually put an end to military dictatorship.[4] Although forceful demolition did not entirely disappear (Choi 2012), poverty came to take a subtle and invisible form with the near-completion of redevelopment processes. As one-time activists transformed themselves into politicians or government officials, the poor's protests against the state have been replaced by so-called public-private partnerships, in which activists engage in community-based projects with financial support from local governments (Cho 2015; Cho and Lee 2017).

---

2 People's Solidarity for Participatory Democracy website, accessed 10 January 2017, http://www.peoplepower21.org/English/37917

3 The term "CO" was not popular until the mid-1990s, when the scope of the "people" that South Korean activists targeted was expanded from *binmin* (poor people) to *jumin* (residents of a certain locale). "CO" began to be used widely when activists tried to coin a new term for "*jumin* movements." Today, activists tend to reconstruct South Korean histories of community activism by universalizing the term. See Cho (2015, 141–3). The names of persons and local institutions (e.g., KONET, KACO, Co-Village, and Peace Village) explored in this chapter are pseudonyms, except for the names of well-known activists (e.g., Je Jeong-gu and Jeong Il-u) and historically traceable organizations (e.g., ACPO, ACHR, LOCOA, FRSN, and UPC).

4 Similar to the way in which torment has become a narrative of the past, Park's chapter in this volume points out how garment workers' ongoing presence is buried under the images of the past.

Political transformations in South Korea have confounded not only the meaning of solidarity but also the nation's changing relationship with other Asian countries. Since the mid-1990s, South Korea's budget for official development assistance (ODA) has increased radically, as the government announced the shift in status from a recipient to a donor nation. Joining the Organization for Economic Co-operation and Development (OECD) in 1996, the country has attempted to declare its "great economic success from the ashes of the Korean War" and demonstrate "how ODA can play a crucial role in overcoming the hurdles of development" (Lee 1997, 1). Today, numerous South Korean students, volunteers, workers at non-governmental organizations (NGOs), religious groups, technicians, entrepreneurs, and government officials head to "underdeveloped" or "less-developed" countries. Most of them define their activities in terms of "international development," not solidarity.[5] The narratives of horizontal comradeship among LOCOA activists are now rarely found in the mission to provide help or assistance to impoverished others.

This chapter examines the globalization of South Korean community activism amid the rearrangement of anti-poverty agenda among Asian countries, as well as the shifting political and social economy within the nation. My emphasis is on showing how South Korean activists have not so much abandoned the seemingly anachronistic slogan of solidarity as tried to reinterpret and revitalize it by remapping poor urban neighbourhoods in Asian countries as "locations of reflexivity" (*seongcharui hyeonjang*).[6] In 2012, KONET members founded the Korean Action for Overseas Community Organization (KACO) in order to bridge community organizing and international development. KACO has invoked reflexivity as a crucial part of its activism. In its scheme, reflexivity is a way of denaturalizing conventional rules and practices among people who work in the realm of international development. It

5 See also Jeong's chapter on foreign aid through marriage migrants' kinship network in this volume.

6 In anthropology, "reflexivity" commonly refers to ethnographers' awareness of their relationship to the field of study. Since the late 1970s, many anthropologists have reflected on both fieldwork and ethnographic writing, questioning how they are saturated with the colonial baggage of their discipline, as well as with the problematic representation of otherness. Reflexivity in activism, which I analyse in this chapter, does not deal with the power relations of knowledge production as seriously as in anthropology. Yet, both parties are resonant with each other in that they extend the scope of interlocutors in the field, problematize the uneven relationship between themselves and ethnographers/activists, and prompt self-reflection and self-criticism through their engagement with others.

is also a way of restoring the CO spirit and values, which South Korean community organizations are thought to have lost in their project-oriented, institutionalized actions. I will probe the workings of reflexivity with an eye toward two ethnographic instances. One is the CO training that KONET members have provided to young workers from development NGOs, where most trainees had once conducted community-based programs in the aid industry. The other instance is overseas training, which has been organized by a social welfare corporation with a long and distinguished history of grass-roots activism in South Korea. KONET and KACO coordinated the corporation's visits to urban poor neighbourhoods in Thailand and Indonesia in tandem with LOCOA.

By delving into the two ethnographic examples, my ultimate purpose is to shed light on the relationship between reflexivity and solidarity in moments when radical actions for resistance are on the ebb and project-based anti-poverty interventions such as aid, welfare, and care have become the dominant approaches to "the poor." To achieve this aim, I explore dialogues between the activists' way of seeing Asia as "locations of reflexivity" and the scholarly focus on Asia as method while unveiling the insights and dilemmas of both. Despite being interpreted slightly differently among scholars, Asia as method is an attempt to provincialize and decolonize the West's epistemological hegemony. Central to this attempt is reflexivity – that is, to reach a new self-consciousness through the examination of "others" (Yoon 2014, 194). Defining Asia as method as "a self-reflexive movement," Kuan-Hsing Chen explains its potential in developing new paths of engagement: "The potential of Asia as method is this: using the idea of Asia as an imaginary anchoring point, societies in Asia can become each other's points of reference, so that the understanding of the self may be transformed, and subjectivity rebuilt" (Chen 2010, 212).

Importantly, the Asia as method scholarship contributes to making it possible to explore self-reflection in relation to the ethics of solidarity. In the shadows of ever-increasing global violence, "the reflecting subject" in Western philosophy has emerged as a crucial theme for interrogating how to ethically undertake the responsibility to help address the failures of modernity. Judith Butler writes, "Critique finds that it cannot go forward without a consideration of how the deliberating subject comes into being and how a deliberating subject might actually live or appropriate a set of forms" (2005, 8). Drawing on Emmanuel Levinas's study of the Other, Butler argues that the ethical preoccupation with the individual self has dangerous implications for legitimizing the elimination of the other. Instead, what she proposes is a theory of subject formation that acknowledges the limits of self-knowledge, which may open up "a possibility for acknowledging a relationality that binds me more deeply to language and to you than I previously knew" (40). In this way, a human appears not as an autonomous self but as a precarious self that is conditioned by relationality ("I am my relation to you"). While Butler's final destination is to discover a living place for "I," the Asia as method scholarship expands the reflecting subject from "I" to "us" by highlighting inter-referencing in the region with an eye to the "locations of reflexivity." These locations enable us to push further our discussion of ethics and responsibility as an opportunity to advance new thinking for solidarity, instead of limiting them to technologies of the self.

By focusing on the formation of locations of reflexivity in globalized community activism in South Korea and forging dialogues between it and the discussion of Asia as method, I will analyse the kinds of predicaments and new thinking the linkage of reflexivity and solidarity has catalysed. South Korean community activism sheds light on insights and tensions embedded in such a linkage in various ways. For instance, such activism is differentiated from the globalization of South Korea's ODA (the so-called Saemaul ODA) in that community activists pursue not the global influence of the nation but rather horizontal solidarity among varied locations in Asia. Nevertheless, it is important to ask *to what extent* can locations of reflexivity give rise to the transformation of "self-consciousness through the other" and thus build up solidarity between locations? As I will detail, the work of mutual referencing is based on historical ignorance as much as on historical awareness among different locations. The comparison between the "present" of one location and the "past" of another runs through such work. Furthermore, this type of work tends to generate an affective turn in activism without interfering with the systemic and institutional changes that have posed a considerable dilemma for it. Despite these limitations, I argue that the elusive linkage between reflexivity and solidarity awaits a new conceptualization of solidarity, opening up new ways of thinking about it. Solidarity is not necessarily limited to the interests that political forces seek when they articulate their demand upon the state in the name of the social (Rose 1996, 329) or attempt to bring systemic change to counter structural violence. It is also captured in a scene where people acquire the power of reflexivity – that is, where we reach some recognition "in which precisely our own opacity to ourselves occasions our capacity to confer a certain kind of recognition to others" (Butler 2005, 41).

In what follows, I will fully detail the aforementioned ethnographic instances of South Korean community activism. Before doing so, however, let me briefly introduce KACO – a group that has pushed forward the globalization of South Korean community activism – by focusing on the trajectory of its founder.

KACO: On the Edge between Community Organizing and International Development[7]

In the mid-1990s, I first met Eun Sil – a founder of KACO – in Bongcheon-dong, a southern area in Seoul where the demolition of shantytowns had been a primary issue. There, she acted as a community activist while I volunteered to take care of children whose parents were busy with anti-eviction struggles. When I interviewed Eun Sil some twenty years later and asked her how she became interested in global poverty issues, she reminded me of that time in the 1990s, when many community activists found themselves in a predicament. As she recalled, her self-identification as an activist had begun to falter during the Asian financial crisis in the late 1990s, when "many activists started merely conducting service delivery" as part of government-sponsored welfare programs. This crisis reached its peak in 2001–5, when she organized the relocation of low-income residents to rental apartments as a result of redevelopment processes. Once they secured new housing, former slum-dwellers who had previously fought together against demolition were scattered about. Newly arriving staff members in community centres were devoid of what she called "the self-consciousness of activism" (*undongjeok maindeu*). The seeming de-politicization of urban communities led Eun Sil to question neoliberalism, not only as a changing mode of capitalism but also as a specific mode of governing people's affect and conduct (Eun Sil, interview with the author, 20 December 2013).

In order to investigate how neoliberalism actually affected local communities, Eun Sil headed to the Philippines, a place she had become acquainted with through LOCOA. As she recounted, she hoped to regain her vigour in the Philippines, a country where grass-roots activists had in the 1970s organized a squatter community of 250,000 in Manila. Yet, what intrigued Eun Sil most during her stay was the presence of development NGOs. Framing their activities in terms of "international development," these NGOs frequently visited local communities:

> Staying in the Philippines for about eight months, I came to know that so many development NGOs in South Korea and elsewhere dispatched volunteers and staff to the country. Witnessing their activities, I really felt that I had found a blueprint for activism under global neoliberalism.

7 Some portions of this section appear in Cho 2015.

> Their acts seemed to model the linkage of the global and the local. However, I was surprised to find that local activists in the Philippines kept complaining about South Korean staff's feelings of superiority as well as their irresponsibility. The gap between South Koreans' appreciation of their own work and the local activists' view of them was remarkable. (Interview with the author, 20 December 2013)

Returning to South Korea in 2007, she began to study the discipline of international development: she joined in various events, workshops, and conferences relating to it while introducing herself to development NGOs. In 2011, together with young members of development NGOs to whom she, as a KONET trainer, had given CO training, she formed the Co-Village, a group formed to discuss people- and community-centred models of international development. For about nine months, she interviewed more than fifty figures in development NGOs, to whom she was introduced by those youths. These interviews made her realize that experts in the realm of international development were unable to produce alternative voices: "Most interviewees disliked the overly nationalistic discourses of Korean ODA. They also criticized the structure of aid projects that made people-centred development almost impossible. Because their funds came mostly from the government, however, they hesitated to voice their opinions publicly. In particular, young employees who had just returned after their dispatched work in 'recipient' countries were afraid of disclosing the problems of their organization despite their serious awareness of them" (interview with the author, 10 February 2017).

Through a series of interviews, visits, and studies, Eun Sil felt compelled to bridge the gap between international development and community organizing, and to implant the ethics and methodology of CO among young, passionate workers in development NGOs. For this purpose, she founded KACO in 2012, in consultation with other members of KONET. Without financial support from the government, KACO was funded by KONET and other CO-related organizations, as well as by progressive development NGOs. Nevertheless, this move was not entirely smooth. At first, Eun Sil had difficulties persuading veteran CO activists of the need for KACO. Reminding her of the long tradition of international solidarity through LOCOA, many activists in KONET wondered why they should make new friendships with those who had worked from the outset in close partnership with the government. However, Eun Sil asserted that interactions through LOCOA had already become nominal and lost vitality.

> Today, South Korea has little common experience to share with other Asian countries. We now witness the apparent differences in poor people's experiences and socio-economic conditions in different countries. Many activists in Indonesia, Cambodia, Laos, Myanmar, and the Philippines also think that way. The poor still resist and fight against state violence there. In Korea, however, governance[8] has become a mainstay of community activism after the forceful demolition has decreased. Most CO activists are busy conducting state-led community programs in their own neighbourhoods. If the annual meeting of LOCOA is coming up, they suddenly gather together and improvize a report to prepare for it. This process makes everyone feel tired and listless. (Interview with the author, 10 February 2017)

Under circumstances where Asian activists see more differences than similarities in each other's locations, what is the motivation for pursuing international dialogues relating to community activism? While creating a new relationship to the realm of international development, how has KACO sought to approach the globalization of community activism differently than its predecessors? Though not publicized explicitly, reflexivity has served as the primary methodology of KACO, as I will outline in the following two sections.

Co-Village: Reflexivity Regarding What We Have Naturalized

As noted, the Co-Village started as a group for (incumbent or former) young workers in development NGOs to share their anxieties about the industry of international development and to discuss people- and community-centred models of international development. Most members had experience conducting development programs in Asia or Africa for two or three years as employees of large-scale development NGOs. After returning to South Korea, they organized a seminar group for studying alternative models of development, and invited Eun Sil, a KONET trainer, to teach them the CO methodology. Frustration about the realm of international development led a small seminar meeting to evolve into a solid group of about fifty members: the group organized regular CO training courses, created an agenda for linking CO to international development, and contributed to the foundation of KACO.

8 Eun Sil here used the term "governance" to indicate an increasingly institutionalized partnership between community organizations and the local government.

For example, returning after being dispatched to work in Laos for two years, Seo U became a member of Co-Village. As she recalled, "Most young staff members, including myself, shared frustration while working in development NGOs. Receiving top-down directions from the headquarters in Korea and implementing them in conflict with native coordinators and villagers, we felt tired and helpless. Senior officials of development NGOs brushed aside our frustration, treating it as a sort of 'rite of passage.' Co-Village provided us with a shelter for sharing our worries. It was a great comfort to us" (interview with the author, 20 May 2015).

Since 2011, new members of Co-Village have received CO training for about three months under the guidance of Eun Sil and other KONET activists. Reflexivity has been central to the training courses. Trainees are expected to look back on what they did in the field during their time in Asia and/or Africa and reconstruct theses sites as "locations of reflexivity." Points of reference for reflexivity include conducting interviews with veteran CO activists and progressive practitioners from development NGOs, reading books about well-known activists in and outside Korea, visiting historic sites of community activism, and participating in memorial events for late activists or in international exchanges with Asian activists. The basic premise of reflexivity referencing is that the locations of community activism cannot be distinguished as being "at home" or "abroad." Whether they come from the records of the past or the present, or from the stories of South Korea or other countries, all have served as locations of reflexivity.

In the fall of 2015, Co-Village members were expected to write a reflective essay on a book entitled *A Tale of Jeong Il-u* (Jeong Il-u iyagi) and discuss it during their training process. Jeong Il-u (John V. Daly, 1935–2014) was a long-esteemed Catholic priest and grass-roots activist. Born in the United States, he eventually settled in South Korea and fostered solidarity among the evicted poor, despite continuous threats of exile under military rule. In 1973, he and Je Jeong-gu (the late South Korean activist and politician, 1944–99) met in Cheonggyecheon, the largest slum area in Seoul.[9] On three occasions between 1977 and 1985, they led collective

---

9 Many factories in Changsin-dong – the focus of Park's chapter in this volume – used to be located in Cheonggyecheon. In 1970, Chun Tae-il, a worker at a Peace Market clothing manufacturer in Cheonggyecheon, committed suicide in protest of harsh working conditions. Factories in Cheonggyecheon gradually moved to Changsin-dong and other nearby areas in the 1970 and 1980s, as Chun's death led to the unionization of workers and prompted the government to enforce some regulations regarding labour protection See N. Han (2017, 34).

migrations for poor people who were evicted from Cheonggyecheon and other shantytowns in Seoul. With the evicted poor, they built "Peace Village" in the outskirts of Seoul. Since then, Peace Village has remained a legendary place of South Korean community activism.[10]

When asked to write about how Jeong's life could lead them to view the locations of international development in different ways, some trainees at Co-Village newly identified Jung as an American who had lived with the poor in a remote country. They thus compared his life with their own lives as staff in the realm of international development: "I enjoyed reading *A Tale of Jeong Il-u*. Like him, I will be a foreigner in my field. Priest Jeong was the very person who put his CO thoughts into action in the field of international development"; "I was impressed by Jeong's humility and desire to become an ordinary resident of Peace Village"; "Jeong made me realize that long-term stays with local people would bring about changes for the community spontaneously." Furthermore, Jeong's "quiet" activism, which took a long time to bear fruit, led some trainees to reflect on their "loud" community projects: "Haven't we destroyed the freedom of local residents by enforcing time-sensitive projects regardless of their will?" "Can we really become not a strange foreigner but a real resident in our locations of international development?" (Co-Village 2015, 268–76).

Through a series of training practices that supplemented workshops, interviews, on-site visits, and reading books, the members of Co-Village transformed their project sites in Asian and African countries into locations of reflexivity. That is, they reflected on these sites as locations where (as shown in their narratives) they had mobilized local residents against their will instead of encouraging their voluntary participation, treated those residents as a kind of tool for achieving project goals, and transplanted a "universal" model for success without considering various political, economic, and cultural differences. For example, Ji Hyeon, a trainee of Co-Village, interviewed a veteran activist who used to run a day-care centre in a shantytown in Seoul. The activist's contrast between "community building through long-term relationships to the poor in the past" and "short-term and performance-oriented projects in the present" prompted Ji Hyeon to de-naturalize her own experiences of conducting development programs in the Philippines: "The activist, who devoted her life to grass-roots activism for three decades, made me brood over the meaning of the word 'speed.' In the Philippines, I took it for granted that

10 In 1986, Je Jeong-gu and Jeong Il-u won the Ramon Magsaysay Award in recognition of their community activism in South Korea.

community programs should be done depending on my own speed, not the speed of local people" (Co-Village 2015, 319).

Such self-reflection did not mean that these Co-Village members decided to leave the industry of international development in search of a more radical mission. Most Co-Village members kept working in development NGOs, although they continued to join events organized by KACO and participated in some political rallies in the name of Co-Village. In this sense, the making of locations of reflexivity through CO training at Co-Village raises some questions. How can the problems of international development, which Co-Village trainees newly discovered, be dealt with when the trainees return to development NGOs and other related agencies as front-line practitioners? If the performance-based, business-like community projects that they denaturalized through reflexivity are structural rather than individually inappropriate, what does it mean to say that front-line workers of this industry desire to identify themselves as community activists? Let me turn to another ethnographic instance that raises a similar problem.

Overseas Training: Reflexivity Regarding What We Have Lost

As discussed above, South Korean community activism has become a subject of learning and respect for those who feel disappointed and exhausted by the standardized system of international development. The CO training has helped them engage seriously with "people" and "communities," which are buzzwords in the realm of international development. Nevertheless, it should be noted that what they commonly call "CO" does not necessarily represent the landscape of present community activism. CO activists are increasingly confronted with the need to fulfil a new role as business operators as they compete to apply for projects sponsored by governments, corporations, churches, and large-scale NGOs. KONET trainers rely heavily on the past experiences of their seniors because they find it difficult, albeit not impossible, to bring up pertinent examples of best practices from current CO activities (Cho 2015, 153–4).

Peace Village is no exception to this trend. It is a community made possible through the collective migration of the urban poor in the 1970s and 1980s. Well-known CO activists Je Jeong-gu and Jeong Il-u purchased land on the outskirts of Seoul and managed the costs of construction through the sponsorship of a German Catholic foundation.[11]

11 A total of 471 households from three slum areas in Seoul moved to Peace Village in 1977–85 (Jeong Il-u 2009, 86).

Je, Jeong, and relocated residents built houses in Peace Village together and, by working in nearby factories, paid back the money they borrowed from the foundation. Running a co-op credit union, producers cooperative, day-care centre, library, and scholarship association, poor people and activists survived amid Korea's rapid modernization drive and military dictatorship. They built a kind of self-contained community, what was termed a "thick mixed rice community" (*jjinhan bibimbap gongdongche*), where "local residents were mixed while they fiercely fought, reconciled, backbit, and praised each other" (Jeong Il-u 2009, 87). As Jeong Il-u wrote in his memoir, "Wherever in rural or urban areas, the poor cannot live alone. They need to live together. I haven't thought of anything except that poor people cannot but live with community" (ibid., 90–1).

When most cities went through the brutal demolition of unlicensed houses and anti-eviction struggles in the 1980s, Peace Village served as the sort of model for the poor community that CO activists in other areas hoped to emulate. However, the working of the community has been rapidly institutionalized since the 1990s. Along with the near-completion of redevelopment processes and the evolution of electoral democracy, activists in Peace Village reframed their CO histories in terms of "welfare" and registered their main centre as the Peace Village Social Welfare Corporation (PV) in 1996. This transformation into a corporation meant that PV came to conduct many welfare-related projects in partnership with the government or to run welfare-related institutions outsourced by the government. Over two decades, the one-time community where poor people and CO activists communalized production, education, and livelihood has dramatically shifted into an ordinary neighbourhood, where initial settlers are heavily outnumbered by new immigrants, and local residents use PV's facilities as welfare clients or customers. In this process, PV has become a sizable corporation under which seven institutions conduct diverse community programs for "at-risk populations" in tandem with government or corporate bodies. Social workers, not community activists, constitute the great majority of PV employees.

Such spillover expansion has been a source of worry among senior members of PV, who still remember the old days in Peace Village. Jung, the director of a community welfare centre affiliated with PV, explained to me the reason why:

> Since the building of PV, we have relied heavily on government subsidies. Our activities have been brought into the regulatory system of the government. Although the founders are certain that the CO principle and spirit should survive under the corporate system, most employees have

> found themselves stuck in bureaucratic documentation. 2011 was the fifteenth year of the founding of PV. At that time, many senior members, including myself, raised voices of self-reflection. Our roots come from the CO principle and spirit. We asked ourselves if we really abided by them. (Interview with the author, 17 February 2017)

PV's overseas training began in this context. With senior advisers, the chair of PV – an early activist in Peace Village and the widow of the late Je Jeong-gu – formulated a new mission for the corporation and sought a way to "re-educate" its employees. For this purpose, senior members tried to find "locations" where people put the mission for revitalizing the CO principle into practice. As Director Jung remarked to me:

> [In Southeast Asia], housing rights are ignored, and evicted people endure social suffering ... Of course, you can find these problems in Korea, but they are made invisible in most cases ... In Korea, you can also find communities where grass-roots activists still struggle to realize the CO methodology. However, we seniors suggested that we go to relatively unknown locations outside Korea, which might be closer to the sort of original form of CO. We thought that this way would be more effective to re-educate our workers. (Interview with the author, 17 February 2017)

To explore the effects of "re-education," twelve senior officials in PV first visited poor urban neighbourhoods in Thailand for six days in May 2013. KACO coordinated PV's visit in conjunction with LOCOA,[12] and a young member from Co-Village joined as an interpreter. Activists in the Four Regions Slum Network (FRSN) in Thailand – a member of LOCOA – guided PV officials to a homeless centre, public land near a canal and railroad tracks, and other slum communities in Thailand, where they were organizing poor people to improve their living conditions, to respond to lawsuits and evictions, and to urge the government to solve land conflicts between slum dwellers and private landowners.[13] Director Jung looked back on her travels, saying that all visitors from PV "gently speaking, learned a lot, and roughly speaking, got shocked." They were "shocked" by the fact that "*homeless* people" (Jung's emphasis) had finally won a long-term land-lease agreement

12 Despite her criticism about "nominal" interactions among Asian activists affiliated with LOCOA, Eun Sil continued to communicate with that group. Yet, she placed more weight on LOCOA's role of linking different CO locations than on formalized meeting among activists.

13 See the 2014 annual record of PV.

from the government through sixteen years of resistance. They were also "shocked" by a scene in which slum dwellers commonly called community activists "our family" or "comrades." Such observations led her to question her own practices:

> Most of us received social work education. Basically, social workers focus on how to allocate government subsidies effectively and how to satisfy social work targets by planning good programs. They tend to use the word "target" (*daesang*) without question. For those who frequently did surveys to find welfare need, community organizers in Thailand seemed to do nothing. It was shocking to see them do nothing while slum dwellers do everything ... We came to realize how impatient we had been in PV – that is, how we couldn't wait for our residents so that they could solve their problems for themselves. I couldn't guess how much work the community organizers had done until local people did it that way. (Interview with the author, 17 February 2017)

Similar responses continued to emerge in subsequent training programs. Acknowledging the benefit of overseas training, PV has expanded this opportunity to low-level officials, most of whom are front-line social workers. Through the coordination of KACO and LOCOA, PV officials received CO training in various communities in Indonesia: fifteen officials visited Surabaya for six days in November 2014, and eighteen officials visited Makassar and Jakarta for six days in August 2015. Most communities were organized by the Urban Poor Consortium (UPC), a coalition of civil society organizations focusing on urban poverty issues.

It is important to note that most CO "locations" that PV officials visited faced tense situations. They seemed to be different from community centres in PV, where social workers implemented routine programs. In the Indonesian locations, for instance, people who lived on the banks of a river were threatened with eviction on the grounds that they polluted the water. Some poor people whose land was purchased at an extremely low price by a big company struggled to recover their land rights. Others were in danger of losing their community as a result of a leakage accident caused by an oil company. They were urged to leave, receiving only meagre compensation, because the accident was framed as a "natural disaster" through an alliance between the company and the local government. Regardless of whether CO activists tried to challenge the irresponsible decision made by the government or prevent the forceful relocation through negotiations based on the electoral power of the poor, every situation was highly intense and urgent.

PV officials were deeply impressed by these desperate actions for, as well as passionate attachment to, these communities. Realizing how they had become inured to regular and banal tasks, these officials were re-awakened by their time in Indonesia and by viewing that site as a location of reflexivity. In the previous section, we saw that Co-Village members reflected on their naturalization of the system of international development. In this section, PV officials brought up the CO principles and spirit, which they thought they had lost, as the target of reflexivity. When asked to give their impressions of CO locations in Indonesia, a PV official compared the present state of poor communities in Indonesia with the past of Peace Village, which she had only heard about from senior members. She said that her visit to Indonesia provided her with an opportunity to imagine how Je Jeong-gu and Jeong Il-u would have communicated with local residents. Another official began to question why her work in PV was not as "touching" as at CO locations in Indonesia.[14] Coordinating PV's overseas training, Eun Sil emphasized that this kind of reflexivity was made possible only through the encounter between "locations."

> In the process of training, it seemed that PV officials felt the "love" (*aeteutham*) of the locations that they visited, although this expression might not be objective. Some officials looked at those whom they met in Indonesia as fellows. Other seniors reminded them of their younger days in Peace Village. Self-reflection and a fellow feeling of love naturally emerged because they did not simply hear stories but faced "locations" (*hyeonjang*) straightly … "In Korea, I'm becoming a machine." "I'm now nothing but a technician obsessed with projects" … By being at the "location," they began to confess what they had silenced for fear of criticism. I think there's no moment as touching as this. It's touching because they let out what they couldn't say due to shame. (Interview with the author, 10 February 2017)

What happened in PV after the overseas training? Jinwoo, a low-level official, told me that social workers tried not to "objectify" (*daesanghwa*) local residents. "No one induced them to, but social workers began to help building various self-help groups among local residents. Some of them made community programs in dialogue with residents, instead of doing it on their own authority" (interview with the author, 17 February 2017). He smiled while saying that senior officials became

14 See the 2014 PV Training Packet.

more generous about calculating the inputs and outputs of welfare programs. After completing the overseas training in Indonesia, a senior official wrote about how community leaders there successfully preserved CO values: "They made us realize what it meant to let people speak for themselves and let them solve problems by themselves.[15] Every moment, we got inspired and challenged by them."[16]

However, one may have noticed the difference between the "CO" brought up by the PV official and the "CO" highlighted in the locations in Indonesia. In the poor communities in Indonesia, the CO principles – stating that community organizers and leaders should believe in people's power, wait for people's initiative, and act together with people – were based on urgent situations, such as violent evictions and forcible relocations. In today's Peace Village, however, the CO principles are interpreted and operate within the boundary of community "projects." In Indonesia, the UPC called for people's "participation" for *survival* – that is, in order for them not to be evicted and deprived of their land rights. Yet, PV requires "participation" as an indicator for measuring the success of a series of community-oriented projects while constituting people as a governable group. In such a relationship of governance, people "must be made to act"; otherwise, they are to be criticized for their "nonparticipation," "powerlessness," and the "lack" of self-esteem (Cruikshank 1999, 82, 83, 93). Like the ethnographic example of Co-Village, PV's overseas training leads us to question the political implication of reflexivity, when affective efforts for recuperating "what was lost" do not necessarily interfere with the shifting relationships of governance.

Dialogues between Community Activism and Asia as Method Scholarship

Thus far, I have examined how grass-roots activists in South Korea have re-read poor urban neighbourhoods in Asia as locations of reflexivity while organizing and coordinating CO training practices in diverse ways. Such pluralistic readings of locations are significant for our understanding of the world, as Mizoguchi Yūzō has asserted. In his book *China as Method* ([1989] 2016), which inspired Kuan-Hsing Chen's notion of Asia as method, Mizoguchi criticizes mainstream

15 Song's chapter in this volume echoes how activists in working-poor neighbourhoods appreciate people's autonomy.

16 See the 2015 annual record of PV, 43.

sinology in Japan for measuring China's degree of progress (or lack thereof) by using the world as the standard. He problematized this method because the "world" was a Eurocentric one with a "fixed and pre-arranged method." He thus argued that a world that takes China as method would be "a world in which China is a constitutive element'" – that is, "a pluralistic world in which Europe is also one of the constitutive elements" (Mizoguchi 2016, 516). Much earlier than Dipesh Chakrabarty (2000), he attempted to provincialize both Europe and China instead of reinforcing the opposition between East and West (Murthy and Sun 2016, 503).

Mizoguchi's insight is exuded in the creation of locations of reflexivity, which I have explored through the two ethnographic examples above. Such creation distinguishes South Korean community activism from the nation's Saemaul ODA, although both of them commonly seek the globalization, or "South-to-South" interaction, of anti-poverty interventions. Since 2011, the South Korean government has attempted to export the Saemaul Undong (New Village Movement), the nation's rural development campaign that took place under military rule in the 1970s, as the representative development model for ODA.[17] The preachers of Saemaul ODA highlight their "non-colonial" and "non-Western" position but nevertheless identify themselves as passengers on the linear trajectory of modernization, as did Korean and Korean-American missionaries in "less-developed" countries (J. Han 2010, 147). In other words, the government discourse addresses the Saemaul ODA's contribution to South-to-South interaction as a source of national pride: it merely raises South Korea to the rank of "the West" without problematizing the conventional principle of modernization.

Importantly, in contrast to the Saemaul ODA, community activism no longer considers the nation state as the primary unit of global interactions. On the occasion of the fourth anniversary of the KACO in July 2016, Eun Sil declared to the audience that "the world is much the same as Bongcheon-dong." In her statement, Bongcheon-dong was neither

17 In fact, Saemaul Undong is traceable to various indigenous movements for rural development preceding the state-led campaign in 1970 (Kim 2009). It has also been interpreted and appropriated by rural people in unorthodox ways (S.-M. Han 2004; Oh 2014). However, such heterogeneous historiography does not deny the fact that the discourse of national development overtook other voices when local flows of rural development were incorporated into the main agenda of the Park Chung-Hee regime. In the shift of Saemaul from a rural development campaign in the 1970s to a foreign aid program in the new millennium, "national development" continued to exert discursive dominance. See H. Jeong (2017).

an administrative place-name in Seoul nor the erstwhile site for the anti-eviction struggles that she had long engaged in during the 1990s. Instead, it emerged as a metaphor of a location (*hyeonjang*), which would become critical through the encounter with another *hyeonjang*. Dissatisfied with "the state" being the first and foremost grid in the regime of international development, Eun Sil tried to conceptualize *hyeonjang* as a suitable alternative: "I like the word *hyeonjang*. I frequently say 'I go to *hyeonjang*' or 'I see *hyeonjang*.' In my 20s and 30s, Bongcheon-dong was my *hyeonjang*. Now, my *hyeonjang* could be somewhere in Surabaya and Makassar" (interview with the author, 10 February 2017).[18]

Indeed, globalized community activism in South Korea resonates with scholarly discussions of Asia as method in interesting ways. Both advance a new imagination by separating the boundary of *hyeonjang* from its national borders. Like Bongcheon-dong in Seoul and a village in Surabaya, the Korean peninsula, Okinawa, Kinmen (Quemoy), or Diaoyu (Senkaku) are considered core locations (McCormack 2011; Sun 2011; Baik 2013a, 2013b). Though marginalized or particularized in the global world order, each location may steer what Baik Young-seo calls interconnected East Asia "away from a New Cold War, and toward an East Asian Community" (2013b, 137).[19]

More importantly, it should be noted that both South Korean activism and Asia as method scholarship shed light on the linkage between reflexivity and solidarity. KACO's attempts to train and spread CO methodology configure Asia's poor neighbourhoods as locations of critical reflexivity, which people learn from and refer to across time and space. They aim at a pluralistic world where no hierarchy among locations exists, no boundary of the nation state matters, and horizontal solidarity emerges through encounters between locations. Asia as method scholarship furthers the method of reflexivity while focusing on the imaginary unit of "Asia" (or "East Asia"): "A society in Asia may be inspired by how other Asian societies deal with problems similar to its own, and thus overcome unproductive anxieties and develop new paths of engagement" (Chen 2010, 212). Such reflexive dialogues between locations ultimately aim at achieving solidarity, whether it is built among (poor) people, activists, or intellectuals. Prompted by "a

---

18 Park's chapter in this volume provides an excellent elaboration of competing meanings of *hyeonjang* in a different context.

19 For thinkers grounded in Asia as method, however, geopolitics based on national borders is still significant because it leads us to better understand how "core location" is doubly marginalized in the hierarchy within East Asia as well as in Eurocentric world history. See Baik (2013a, 17).

self-reflexive movement" (ibid., 213), this solidarity is full of affect. In CO training programs, participants were "moved," "shocked," or felt "love" through the encounter of locations. Baik Young-seo proposes "co-suffering" (*gonggo*) – that is, sharing suffering – as an affective condition for solidarity (Baik 2013a, 26–7).

Yet, in what ways and to what extent do the locations of reflexivity lead to solidarity? As mentioned in the introduction to this collection, notions of core location and Asia as method have been developed by humanities scholars. In their approach, the relationship between reflexivity and solidarity remains elusive: their mission is to cast some new direction for rethinking the global order rather than to examine how the direction is actually performed in practice. Yet such elusiveness may appear problematic for activists who seek social change through their actions. Indeed, South Korean community activism has actively incorporated what Kuan-Hsing Chen (2010, 213) describes as the "political motive of Asia as method" – that is, "the use of Asia as an emotional signifier to call for regional integration and solidarity." However, what if this work of mutual referencing leads to an affective turn in activism, without addressing the systemic and institutional issues that have created the initial dilemmas for it? How is it possible for the affective locations of reflexivity to lead to a sort of solidarity that both South Korean community activism and the Asia as method scholarship strive for?

As I have shown in the previous two sections, CO training practices organized by community activists have rarely brought about immediate and visible changes in the lives of trainees. Young members at Co-Village felt that, under the present circumstances – according to which they were expected to adjust to the short-term cycle of development projects – the CO-centred movement was something remote from them. The more they were involved in CO training, the more they felt that people-centred development was impossible, short of leaving their current positions. Although trainers emphasized the possibility of people fighting against unjust powers in their own communities, most of the development NGOs to which they belonged tried to avoid taking an ideological stance (Cho 2015, 154–5). Self-reflection led some Co-Village members to live as a kind of "double-agent." At one level, they, as front-line workers of development NGOs, continued to implement conventional programs, following the ruling office's directions. At another level, they, as Co-Village members, occasionally organized signature campaigns against the malpractices of the nation's aid policy[20]

20 In January 2017, it was reported that Choe Sunsil, a person at the centre of a political scandal involving the impeached President Park Geun-Hye, gained illicit profits from the South Korean ODA program in Myanmar.

or the poor's forceful eviction caused by global capital in "recipient" countries.

PV officials have continued to engage in welfare programs while trying to adjust them to an evaluation index. Some of them were "shocked" by the desperate struggles that they had witnessed in the poor communities in Thailand or Indonesia. Others started to wonder if the scene of violent repression that they witnessed there could also be found in their own country, and they reflected on themselves, those who used to pursue the "improvement" of the lives of the poor in South Korea. Nevertheless, such self-reflection did not enable them to resist the giant system of the welfare industry. After returning from overseas training, PV social workers tried to revitalize the CO methodology, which they believed played a crucial role in the formation of Peace Village in the 1970s and 1980s. As I noted earlier, however, community organizing has become a desperate mission, intended more for the social workers who are expected to boost people's "participation" and "empowerment" than for local residents whose life concerns are not necessarily bounded by their "community."

All in all, the creation of locations of reflexivity contains the danger of instrumentalization – that is, a danger of referencing each other based on each other's need while streamlining and simplifying the particular historical specificities of each location. This critique may also acquire currency in the case of the Asia as method scholarship if it focuses solely on "Asia's rich multiplicity and heterogeneity" against the binary opposition between East and West (Chen 2010, 215). It is important to be reminded that the task of inter-referencing is not external to the uneven power dynamics within and among locations. Without any consideration of those dynamics, the inter-referencing of the CO methodology would mean that it merely shifts from a weapon through which to struggle against state and corporate violence to a means for measuring the "empowerment" of the weak. Nevertheless, does this critique lead us to the conclusion that collective efforts for creating the locations of reflexivity are nothing but "incomplete" and "fictitious"? Rather than entirely dismissing such efforts on account of ignoring the structural unevenness among locations, I conclude this chapter by demonstrating that affective activism, shown in solidarity based on reflexivity, causes us to await action rather than stifle it.

Conclusion

In this chapter, I have examined the emergence of locations of reflexivity in Asia by focusing on two kinds of CO training practices organized and coordinated by grass-roots activists in South Korea. One is

CO training at Co-Village, aimed at young workers from development NGOs who conduct community-based programs mostly in Asian countries. The other is overseas training at poor communities in Thailand and Indonesia, aimed at officials in a social welfare corporation with a long and distinguished history of grass-roots activism. In both cases, CO sites across different regions in Asia were given new attention as locations of reflexivity. Diverse methods of CO education across time and space made participants realize what they had assumed or lost. Rethinking the locations in which they had worked or visited, Co-Village members de-naturalized the apparatus of international development while PV officials tried to revitalize the CO principle in their workplaces.

It is significant that those who construct the locations of reflexivity have not so much dismissed the seemingly outdated slogan of solidarity as pursued it with affective engagement. The prevalence of reflexivity as an ethics of solidarity indicates an affective turn in activism, in which affective dialogues for sharing social suffering outweigh a teleological mission to complete a goal. Emotionally engaged with horizontal solidarity, veteran activists, poor people, NGO practitioners, and social welfare officials encounter one another across time and space. Such encounters lead South Korean participants to realize the limits of their self-understanding and to problematize the techno-politics of global anti-poverty interventions in which structural problems are redefined in technical language and considered easily solvable. Although such problematization rarely leads to immediate action, it still lies in the affective sentiments of those who remember intimacies, passions, and warm camaraderie in these locations. In this sense, the location of reflexivity calls for new thinking about solidarity in order to entail not just direct, immediate action but also the promise of action, which may endure through affective bondage.

When I interviewed Ms. Jung, the PV official whom I mentioned earlier, she tried to share her worries with me. In Jung 2016, four activists from Indonesia visited "CO locations" in South Korea at the invitation of the Je Jeong-gu Foundation. This trip was a kind of "overseas training" for Indonesian activists. Jung was delighted to meet them again and to have an opportunity to take them around "fancy" institutions in PV. Yet, she said, she was also anxious because she quickly realized that they were not that interested in visiting those institutions. As she lamented, what UPC activists really wanted to see no longer existed:

> Because they were struggling to help the re-location of poor people, the UPC activists wanted to know more about Peace Village in the 1970–80s – that is, how activists had negotiated with the government, how they had

> persuaded poor people to come here, how they had collected funding for this task, and so on. They didn't look interested in today's PV, except for a few co-ops. But I felt ashamed to say that even the co-ops received government subsidies. In Indonesia, we really felt touched by what they did. Here, we also want to provide them with some "touching" moments. But what will be these moments? How can we impress them through our contemporary activities, not memoirs of the past? (Interview with the author, 17 February 2017)

Indeed, Jung's story reveals the structural and historical unevenness among locations in Asia, which cannot be dealt with through reflexive dialogues or mutual referencing. Promptly linking reflexivity to solidarity, inter-referencing may lead to mutual ignorance as well as mutual imagination, failing to differentiate itself from liberal pluralism. Nevertheless, I would emphasize that Jung's deliberate questions that problematized state-sponsored CO practices in South Korea might not have emerged without these very encounters between locations. The encounter should be taken seriously because it generates affective uneasiness – the feeling of being "touched" and "ashamed" – and thus urges community participants to think or act differently. In this way, Jung's self-reflection takes on a futuristic form of solidarity, implicitly avowing her promise to be accountable for herself and others.

REFERENCES

Alinsky, Saul D. 1989. *Rules for Radicals: A Practical Primer for Realistic Radicals.* New York: Vintage Books.

Baik, Young-seo. 2013a. *Haeksimhyeonjang-eseo Dong-asiareul Dasi Mutda: Gongsaengsahoereul Wihan Silcheon-gwaje* [Rethinking East Asia in Core Locations: A Task for a Co-Habitation]. Paju, KR: Changbi.

Baik, Young-seo. 2013b. "An Interconnected East Asia and the Korean Peninsula as a Problematic: 20 Years of Discourse and Solidarity Movements." *Concepts and Contexts in East Asia* 2: 133–66.

Butler, Judith. 2005. *Giving an Account of Oneself.* New York: Fordham University Press.

Chakrabarty, Dipesh. 2000. *Provincializing Europe: Postcolonial Thought and Historical Difference.* Princeton, NJ: Princeton University Press.

Chen, Kuan-Hsing. 2010. *Asia as Method: Toward Deimperialization.* Durham, NC: Duke University Press.

Cho, Mun Young. 2015. "Orchestrating Time: The Evolving Landscapes of Grassroots Activism in Neoliberal South Korea." *Senri Ethnological Studies* 91: 141–59.

Cho, Mun Young, and Seung Cheol Lee. 2017. "'Sahoe'-ui Wigiwa 'Sahoejeogin Geos'-ui Beomnam: Han-gukgwa Junggukui Sahoegeonseol Peurojekteu-e Gwanhan Sogo." [The Crisis of "Society" and the Explosion of "the Social": Social Construction Projects in South Korea and China]. *Gyeongje-wa Sahoe* [Economy and Society] 13: 100–46.

Choi, In-gi. 2012. *Gananui Sidae* [The Age of Poverty]. Paju, KR: Dongnyeok.

Co-Village. 2015. *Jumin-gwa Hamkke! Jiyeokgwa Hamkke!* [With People! With Community!]. Seoul: Co-Village.

Cruikshank, Barbara. 1999. *The Will to Empower: Democratic Citizens and Other Subjects.* Ithaca, NY: Cornell University Press.

Han, Ju Hui Judy. 2010. "If You Don't Work, You Don't Eat: Evangelizing Development in Africa." In *New Millennium South Korea: Neoliberal Capitalism and Transnational Movement*, edited by Jesook Song, 142–58. New York: Routledge.

Han, Na. 2017. "Changsin-dong Maeuldolbom-ui Yeogsajeog Jaeguseong" [Historical Reconstruction of Local Community Care in Changsin]. MA thesis, Yonsei University.

Han, Seung-Mi. 2004. "The New Community Movement: Park Chung Hee and the Making of State Populism in Korea." *Pacific Affairs* 77 (1): 69–93.

Inamoto, Etsujo. 2011. *Kkeuchi Eomneun Iyagi* [Never Ending Story]. Seoul: Je Jeonggu Ginyeom Sa-eophoe.

Jeong, Hyeseon. 2017. "Globalizing a Rural Past: The Conjunction of International Development Aid and South Korea's Dictational Legacy." *Geoforum* 86: 160–8.

Jeong, Il-u. 2009. *Jeong Il-u Iyagi* [A Tale of Jeong Il-u]. Seoul: Je Jeonggu Ginyeom Sa-eophoe.

Kim, Young-Mi. 2009. *Geudeurui Saemaeul Undong* [New Village Movement of Their Own]. Seoul: Pureun Yeoksa.

KONET (Han-guk Jumin Undong Jeongbo Gyoyugwon). 2010. *Jumin Undong-ui Him: Jojikhwa* [Community Organization Methodology]. Seoul: Je Jeonggu Ginyeom Sa-eophoe.

Lee, Ho-Chul. 1997. "Korea's Efforts in Official Development Assistance." *ERI Working Paper* 54: 1–30.

McCormack, Gavan. 2011. "Senkaku/Diaoyuui Yeoksawa Jiri" [History and Geography of Senkaku/Diaoyu]. *Creation and Critique* 151: 57–73.

Mizoguchi, Yūzō. 2016. "China as Method." Translated by Viren Murthy. *Inter-Asia Cultural Studies* 17 (4): 513–18.

Mizoguchi, Yūzō. [1989] 2016. *Bangbeob-euloseoui Jung-gug* [China as Method]. Translated by Seo Gwangdeok and Choe Jeongseop. Busan, KR: Sanzini. Murthy, Viren, and Ge Sun. 2016.

Murthy, Viren, and Ge Sun. 2016. "Introduction: The 'Vectors' of Chinese History." *Inter-Asia Cultural Studies* 17 (4): 497–512.

Oh, Yu-seok. 2014. *Bak Jeonghui Sidae-ui Saemaeul Undong* [New Village Movement in the Age of Park Chung-Hee]. Paju, KR: Hanul.

Rose, Nikolas. 1996. "The Death of the Social? Re-figuring the Territory of Government." *Economy and Society* 25 (3): 327–56.

Sun, Ge. 2011. "Minjung Sigakgwa Minjung Yeondae." [People's Perspective and People's Solidarity]. Translated by Seonhyuk Jeong. *Creation and Critique* 151: 74–94.

Sun, Ge. 2013. "Discourse of East Asia and Narratives of Human History." Translated by Gu Li. *International Critical Thought* 3 (2): 215–23.

Yoon, Yeo-il. 2014. *Sasang-ui Beonyeok* [Translating Thoughts]. Goyang, KR: Hyunamsa.

6 The Education Welfare Project at Pine Tree Hill: A Core Location to Assess Distributional and Transitional Forms of Justice

JESOOK SONG

It is amazing to see the change in the way these [government and private welfare agency] people treat us. Who could have known this kind of public welfare [e.g., basic medical treatment, free meals, places to sleep, clothes] would someday be provided? Seeing these changes has made my life "worthwhile" *(sesang cham salgo bol-il ida)*.

– A homeless woman in the Seoul Train Station Plaza, 1999

I sometimes wonder if a few decades' collective efforts to build some autonomy into the community's daily lives (*jagi salmui juini doeneun*) was all in vain. Under the good years of lefty regimes, people with narrow minds deeply influenced by ideological censorship from the time of the military regime hid their thoughts. Now they've become vocal and expose their true thoughts. It negatively affects community activities now, dampening the village atmosphere of decision making through dialogue and cooperation for the common good. It feels almost as if we are back living in the [Korean] War or a refugee camp without any room to appreciate the value of mutual aid and collective action. Without knowing what's going to happen tomorrow, the only priority is to keep surviving in the short term.

– Domin, a community activist of Pine Tree Hill, 2016

This chapter examines the Education Welfare Project (EWP), a school welfare program implemented in a metropolitan working-poor neighbourhood, as a core location of distributional justice.[1] Distributional justice, represented by welfare states, seeks to repair polarized social relations stemming from uneven and structural wealth accumulation

1 The arguments in this chapter will be developed in more detail in my next book.

through promulgating a compensatory system. This is similar to the ways in which transitional justice endeavours to redress historical wrongdoings that cannot be dealt with through the regular court system, by establishing truth and reconciliation commissions and war criminal courts. As the introduction to this volume more fully elucidates, the concept of "core location," as formulated by Baik Youngseo (2013), describes a place that has experienced dual marginalities in geohistory and compound genealogies of praxis that foster insights for a transformative politics of decolonization and anti-capitalism. The Education Welfare Project marks dual marginalities of distributional justice in that it reveals the project's peripherialized location within the capitalist development process; at the same time, the EWP's realization of transformative social relations is impeded despite – indeed, because of – the focus on a certain kind of redistribution as a mediation without problematizing the ways in which the mode of production is systematically buttressing capitalist accumulation. By focusing on these dual marginalities, which cannot be reduced to mere victimhood, this chapter proposes critical approaches to transitional justice as a means of understanding the place of welfare states within the capitalist social totality.

The Asian Financial Crisis and First Welfare States

Since the Asian financial crisis, South Korean urban poor communities have received an unprecedented amount of attention from federal and municipal governments. The new policies have stressed the welfare of its citizens and have attempted to alleviate the social tensions associated with class polarization and poverty that have resulted from the state's single-minded pursuit of economic growth.[2] The period of official

2 Industrial capitalist accumulation in South Korea had already experienced crises multiple times as a result of global influences, such as the oil shock in the 1970s and the electoral democracy achieved by a worker and middle-class alliance in the late 1980s. National growth did not stop after the Asian financial crisis. National industrial development rapidly adjusted its focus to the domain of information and communication technology by promoting a flexible labour market and entrepreneurship (see Seo 2011 and Park's chapter in this volume), which resulted in the magnification of class polarization along with increasing volatility in the real estate market and the exposure of individual households to the global financial market (Jang 2011; K.-K. Lee 2011; Shin 2011). The Asian financial crisis, having taken place at the height of national growth, is therefore more relevant to the perspectives of the working (poor) class who did not benefit from the same portion of the national growth as they had previously.

crisis (1997–2001) coincides with the first appearance of a universalistic welfare state that guaranteed all citizens a basic standard of living. It began with "productive welfarism" (*saengsanjeok bokji*) under the Kim Dae-Jung presidential regime (1998–2003), when the homeless were treated as citizens deserving of welfare, initially as emergency subjects and later gaining permanent entitlements. Productive welfarism was followed by "participatory welfarism" (*chamyeo bokji*) under the Roh Moo-Hyun regime (2003–8), when the Education Welfare Project was first launched to help alleviate class polarization and poverty by designating urban poor communities as priority zones.[3]

The words in the first epigraph of this chapter are those of a South Korean homeless woman who was astonished by the degree of the state's attention given to homeless people at the height of the Asian financial crisis in 1999. The emergence of the gendered homeless subject that made homeless women invisible during the Asian financial crisis is symptomatic of a broader social discourse and ideology fraught with assumptions about gender and class. Homeless men were recognized as former breadwinners of the normative middle-class family and, thus, deserving of state support during the crisis. By contrast, homeless women were invisible and unimaginable: they were painted as unethical and selfish for having left their families and for not fulfilling their motherly and wifely obligations in dire times. Because homeless women living on the street rarely go to public spaces for fear of sexual violence, social workers denied their existence, even when they were standing right in front of them – even in the case of the homeless women in the epigraph (see Jesook Song 2009).[4]

Regardless of the morally laden "invisibility" of homeless women in the public eye, the quoted homeless woman's astonishment regarding the elevated attention given to homeless people reflects a palpable sign of the emergence of the first welfare state in South Korea, "first" in that it claims the universal right of all citizens to a basic standard of living. The universal welfare state emerged under the Kim Dae-jung regime as a national response to the Asian financial crisis. It had unprecedented support from the non-government sectors and dissents groups, owing to Kim's legacy of opposition during the military regimes of the 1970s and 1980s. Homeless people were initially targeted as

3 See Abelmann, Choi, and Park 2012 and Park and Abelmann 2004 for the context of the education crisis; see Jesook Song 2017 for background on the EWP.

4 Throughout this chapter, the notion of the state is divided into ethnographic identification of federal and municipal governments and a conceptual discussion of state sovereignty's role in capitalist political economy.

temporary emergency subjects in distress during the crisis, showcasing the unprecedented benevolence of the welfare state, and became permanently entitled subjects of the state welfare system in 2003 through the Ordinance of Facilities for Protecting Rootless and Homeless People. However, the change in homeless people's status from temporary emergency welfare subjects to permanent entitlement recipients was made possible through the strenuous efforts and lobbying of activists, and by public support more generally (S. Kim 2001; U. Hwang 2007).

Although the mainstream media and civic organizations perceived the idea that homeless people were deserving of state welfare as noble when their new legal status was belatedly implemented in 2005, the ordinance acted as a tool for decentralizing state responsibilities. In other words, the ordinance was set by the federal government in the name of protecting and nurturing those who were vulnerable across the whole population; at the same time, the ordinance became institutionalized through the central government's delegating financial and administrative responsibilities to the municipalities, which do not necessarily possess the resources to operate the programs warranted by the ordinance. The decentralization of the homelessness policy incapacitated the infrastructure building and operability of the majority of municipalities with homeless populations, save for a couple of the largest metropolitan city governments.[5] Some scholars refer to this kind of downloading of the central state's responsibilities to municipalities as a key characteristic of neoliberalism. Rather than assessing the extent to which the South Korean case speaks to neoliberalism, or charting its different trajectory from those of welfare states from (western and northern) Europe and (North) America, this chapter focuses on the ways in which the initiatives of distributional justice are sure to be infelicitous. Such a result is not due to some unpredictable mishap that produced a discrepancy between the ideas in the policy and their implementation on the ground. Rather, I argue that the failure of these initiatives stems from liberal (capitalist and anti-communist) political economic ideology as a crucial condition that engenders a preoccupation with welfare politics, as if there are no other options.

This trajectory of homelessness is not an isolated case of an infelicitous mode of justice seeking within welfare-governing practices in South Korea. The EWP, a school welfare program that mandates utilizing and consolidating the infrastructure of community mutual aid

5 See municipal workers' complaints in Kim Sang-chung's *What I Would Like to Know* (Geu geosi algo sipda), Seoul Broadcasting System, 26 July 2008, episode no. 678.

resources, offers another window through which to understand the intrinsic limitations of distributional justice, which is the core location this chapter hinges upon.

The Education Welfare Project and Children's Network

The extraordinary attention that federal and municipal governments paid to the welfare of urban poor neighbourhoods mobilized grass-roots organizations and community activists to take part in administering social development projects, which stemmed from a yearning to make structural changes. However, the EWP, which aims to redress historical injustices in terms of the class gap under the state's direction, inevitably invites clashes and competition among neighbourhood stakeholders. These clashes and competitions also affect the left in the form of a revitalized censorship in people's daily lives. Tensions rise among local actors, between the municipality in need of actualizing the welfare state's imperative of social development and the local community's attempt to maintain its self-governance. Further, tensions also emerge among community members themselves.

Pine Tree Hill is one of the metropolitan working-poor neighbourhoods affected by the new welfare initiatives, a place I have been frequenting over the past decade or so, tracing the Education Welfare Project. The neighbourhood was designated as one of the top-priority zones of the EWP since the early years of the new millennium. The constituency of the project includes not only the central government's goals of fostering social development and[6] urban regeneration programs. Some of these programs have been launched primarily by municipalities (as Park's chapter in this volume reveals); others are initiated by local grass-roots entities, such as the Children's Network at Pine Tree Hill, which are introduced in the following pages.[7]

The Pine Tree Hill neighbourhood's current population is approximately 27,000. The community still has long-term residents from the 1960s and 1970s, when it was established for relocated post–Korean War refugees, but people have increasingly settled in the area in the

6 The central government's imposition parallels economic development as a form of the state's "selective spatiality" that Oh's chapter precisely contextualizes, because the project is designated as a program for poverty zones to prioritize social development.

7 The names of research participants, neighbourhoods, and organizations – such as Domin, Pine Tree Hill, Seagull Town, and the Children's Network – are all pseudonyms used to protect their identities.

last two decades. Pine Tree Hill residents include taxi drivers, factory workers, low-level office workers, some college students, and people in between irregular jobs. A substantial number of people do not have paid jobs. The latter eke out a living through a combination of hustling, part-time jobs, working as unpaid domestic-care volunteers, and relying on social networks or welfare subsidies. Residents who are eligible for state welfare subsidies earn entitlements based on the following categories: people with disabilities; children, youth, and the elderly receiving no support from family members; people living below the poverty level (*bin-gon gyecheung*); people living just above the basic standard of living, labelled the "lowest income bracket" (*chasang-wi gyecheung*)[8]; people with the status of North Korean refugees; marriage-migrant families; and ethnic Korean returnees under a special decree of reconciliation in response to their forced migration from Korea (e.g., Koreans who migrated to the former Soviet Union or China). The neighbourhood is now one of the most heavily concentrated districts in the city and the nation, not only based on measures of population receiving welfare subsidies, but also in terms of all kinds of "welfare centres" (*bokjigwan*) and "local community centres" (*jiyeok senteo*) targeting "populations at risk," such as youth, elders, marriage-migrant families, people at risk of suicide, and divorced families.[9]

The Children's Network in Pine Tree Hill was established in the late 1990s as an ad hoc association among local grass-roots organizations – including libraries, faith groups, welfare centres, and community centres providing services for children and youth. Although the Children's Network did not emerge merely to support the operation of the Education Welfare Project, it was not necessarily external to the EWP. Rather than existing in parallel with the EWP, the network functioned as key neighbourhood infrastructure in the execution of the Education Welfare Project. Despite the fact that the EWP is promulgated by the central government, primary with financial support, and managed by the municipality, which has administrative responsibilities, including liability as employer, it cannot function without its local infrastructure. Many EWP social workers note that the project was designed as a form of community welfare, although it is labelled as education welfare and anchored

8 See the National Basic Living Security Act, article 2.10, which notes that "The term 'next lowest income bracket' means the low-middle income class, the members of which are ineligible recipients (excluding persons who are deemed eligible recipients pursuant to Article 14–2), and whose amount of recognized income is below the criteria prescribed by Presidential Decree."

9 See Choo 2016; E. Kim 2017; and H. Park 2011 for socially vulnerable goups.

in the public school system. This means that community resources are essential, as they provide the content and the labour that has to be constantly mobilized to execute the after-school or summer programs that the project sponsors. In this regard, the Children's Network is not just an external civic partner to the EWP. Rather, the project is an umbrella initiative that requires both governmental infrastructure (that is, the regular state workers of the municipality's managerial workforce, and irregular state workers of the project's social workers) and non-governmental infrastructure. One of the primary responsibilities of the school social workers in the project is to build and strengthen the infrastructure of grass-roots resources if the infrastructure is not formalized and activated efficiently. EWP workers are contracted state workers with a wide spectrum of educational backgrounds and qualifications, and sometimes there is an overlap between the Children's Network members and Education Welfare Project workers. Most of all, as noted above, EWP social workers are responsible for doing the legwork of building the infrastructure of community resources, not just for the local community's benefit but also for operating the project on behalf of, and under the supervision of, regular government workers, including education board members, school principals and teachers, and regular municipal workers (see Jesook Song 2017). Observers might associate such an approach with a common (neo)liberal practice of post–Asian financial crisis government initiatives in the name of "cooperation between the government and non-government" (*min-gwan hyeomnyeok*). However, the Children's Network was neither independent of the government nor subservient to it, unlike many initiatives undertaken in the name of cooperation. Yet, the network emerged concurrently with the EWP, sometimes sharing initiatives and involving people doing both paid and voluntary labour, and other times instigating projects not involving the government's financial sponsorship.

The Children's Network's mission statement is to help members of the "shantytown" community make ends meet and to foster people's sovereignty in their daily lives (*jagi salmui juini doeneun*). This goal is not confined to Korean urban community organizing (CO). Cho's chapter in this volume succinctly demonstrates the ways in which experienced Korean anti-poverty activists have been inspired by witnessing the successes of anti-eviction movements in "aided" regions in less affluent countries as the essence of CO activism: "They [aided regional CO activists] made us [South Korean CO activists] realize *what it meant to let people speak for themselves and let them solve problems by themselves*" (emphasis added). If Cho's chapter provides a lens through which to witness long-term and more systemized CO activism, the Children's

Network is relatively recent, and the activists' passion for people's sovereignty is full of energy and vigour to the point that community organizers do not necessarily feel the need to follow the CO manuals that are made available to them. The following are a few examples of Children's Network campaigns that were pursued without the government's financial support but that became crucial resources for the Education Welfare Project: the network took the initiative on successful campaigns to serve free meals to children (mostly those attending schools in the neighbourhood who cannot afford to bring a lunch); built a community children's library; decorated village walls with child-friendly drawings; created a parents' and grandparents' group for reading books to children; and supported adolescents' projects of rewriting school textbooks to include critical views on gender and sexuality.

One of the proudest collective memories of Children's Network activists involves a situation in which local people brought pressure to bear against a daily newspaper's misrepresentation of Pine Tree Hill, which it had characterized as a model of "building villages" (*maeul mandeulgi*). Building villages is a decades-old trend in urban revitalization (*dosi jaesaeng*) movements. It highlights residents' self-initiated improvement actions in organizing institutions such as cooperative day cares or alternative schools. It is distinct, though, from the Saemaul Undong (New Village Movement), associated with the military regime's rural development projects during the 1970s,[10] in which building villages has increasingly been dominated by developers and city planners for gentrification and branding municipalities.[11] When a politically conservative mainstream daily newspaper reported that Pine Tree Hill was a successful case of village building, it highlighted the village's previous status as a "shantytown" with broken families and abandoned children. The special report on the village used an image of village children playing computer games as a sign of the pitiful state of children cared for by no family members or neighbours as a result of poverty and divorced parents. The Children's Network was at the forefront of mobilizing Pine Tree Hill residents to demand a public apology from the newspaper. It argued that the paper appropriated Pine Tree Hill as a "building village," something more in the realm of middle-class citizens who could afford to send their children to alternative private schools, and that the

10 See Jeong's chapter in this volume for a critical view of Saemaul Undong's history and its recent revival in foreign aid projects.

11 Regarding municipalities' branding exercises, see also Eom's chapter in this volume about "the Chinatown" project in Incheon and Park's chapter about urban regeneration promotion by Seoul City in the Dongdaemun area.

writer condescendingly represented Pine Tree Hill as a "shantytown." The protest was an action to establish greater sovereignty in their daily lives (*jagi salm-ui ju-in-i doeneun*) against the patronizing representations of mainstream media that overlook the structural and historical problems deriving from class relations and instead frame the problems facing the working poor in moral terms.

The mainstream media might not be a key institution-building actor in social development, but it proliferates the discourse that taps into populist sentiments. When populist politics wanted to see the direction of development in social domains as a marker of democracy or progress, the Pine Tree Hill residents mobilized themselves and refused to be ventriloquized: they refused to be seen and celebrated as a successful case of social development in a manner predicated on representations of their town as one formerly in abject poverty. They recognized how their stories were serving as a mechanism to silence their actual demands for structural changes in social relations – demands requiring a much longer process than any single event of visibility and gesture of redress (Morris 2012).

As much as populist politics wants to appropriate the Pine Tree Hill case, the welfare state is its unabashed agent and has the goal of mediating poverty via the means of distribution without enabling the possibility of eradicating the root cause of inequality – that is, the class contradiction inherent to the capital accumulation process.[12] It was not always clear to community activists and Education Welfare Project workers whether the welfare state's expansion was compatible with the ways in which they would like to build people's sovereignty. For example, since the municipality significantly expanded the EWP priority zones (districts designated as working-poor neighbourhoods by the state) for a decade or so, two outstanding zones were recognized by EWP workers and local community activists: Pine Tree Hill and Seagull Town. Although these zones are the poorest districts in the municipality, Seagull Town was closer to downtown and had undergone recent development in the form of a concentration of high-rise condominiums, whereas Pine Tree Hill was removed from full-scale redevelopment.

---

12 Following Sanyal (2007), Chatterjee (2011) argues that the welfare state works to reverse the effects of primitive accumulation. However, neither scholar situates welfare state history within capitalist accumulation, as not only complementary and contributory but also impeding fundamental change with respect to capitalism. See Donzelot's (1984, 1988) and Castel's (2002) elaboration of the welfare state's history of appropriating the solidarity into the socialization of insurability, as well as Adnan (2015, 35–6) on the limitations of Sanyal and Chatterjee's conceptualization.

A former EWP worker who left the job after a decade devoted to the unionization of EWP workers and building a community network for children and youth[13] claimed that Seagull Town's Education Welfare Project is successful in terms of effectiveness and productivity as a result of its well-planned operation by highly organized and educated social workers. Pine Tree Hill's Education Welfare Project, by contrast, has a strong history of mobilizing for fundraising and building a community library, an exuberant "no-hunger for our children" campaign, and solid community activism.

Leading EWP workers in both zones are former student activists (especially among the older cohorts) or have become activists as a result of their own experiences with precarious labour (some older cohorts and the majority of younger cohorts). These workers are acting together with local residents and community leaders to challenge municipal and state government authorities in the following ways: countering budget cuts for vouchers or non-school performance-related extracurricular activities; transforming their annual contract jobs into permanent ones; and voting for class-conscious candidates for the Education Board. The two competing modes of Education Welfare Project operation – the project planning and management-oriented model in Seagull Town versus the organically mobilized community initiative model in Pine Tree Hill – pursue different strategies to achieve the same goals: people's sovereignty and genuine solidarity that is not subsumed under the state agenda and dictates of capitalist accumulation.

Cold War Legacy and People's Sovereignty

A more pressing barrier to people practising sovereignty stems from the geohistorical baggage of the Cold War. Domin, a charismatic local activist and long-term resident of Pine Tree Hill, was baffled by village leaders' sudden suspicions of her after two decades of rapport and trust between herself and the residents. She was not oblivious to the nation's charged history, which was still ingrained in people's collective memory, where socialists and communist sympathizers were viewed as threats to national security and therefore were subjected to explicit persecution by Cold War military regimes. Nevertheless, people who had provided support during the anti-state civic activism or with fundraising for and then building the local library during the Roh Moo-Hyun

13 Only certain kinds of children were eligible for Education Welfare Projects benefits. For example, high-school students and youths who had dropped out of school were not eligible.

regime were becoming increasingly apprehensive about her activities during recent presidential regimes (i.e., those of Lee Myung-Bak and Park Geun-Hye). To Domin's alarm, they expressed their concern about her activism in accusatory and derogatory terms, referring to her as a "commie" (*ppalgaeng-i*):

> The local community is being heavily influenced by the political scene. I came to think that neither people's consciousness nor the society would change if we just focused on taking care of each other. Our community activism needs to aim at transforming the social structure. When I provided a bit of a different opinion on things like the sexual violence on Shin-an island,[14] our village leaders said behind my back that my ideology is suspicious. Yet they do not say anything in front of me, just smile because of the Children's Network's organizational power [*jojingnyeok*].
>
> They are basically apprehensive about the fact that the Children's Network's political inclination is left-leaning, "North Korea sympathiser-kind-of leftist [*jongbuk jwapa*]." Once I had to challenge them, asking why it was suspicious when I said something very similar to what they [village leaders] said. Then, they said it's because I am a commie [*ppalgaeng-i*]. Those moments have erupted more frequently in recent years. I usually just say "you guys are vulgar." We talk about this by laughing because we have worked together long enough and now we are no longer young.
>
> But to people with whom I have more amicable relationships, I ask, "Have you seen me acting like a commie [*ppalgaeng-i jit*]? Probably not. I'd do more 'commie acts' if all the hard work I've done for the [Pine Tree Hill] village is considered commie." But this reveals that the political scene has changed. In the Roh Moo-Hyun era, people praised our deeds as advanced and ahead of their time. But under the Park Geun-Hye regime political suffocation is more apparent, and the same people who praised us are questioning us saying, "Aren't they North Korea sympathisers [*jongbuk*]"? More and more people brazenly comment on the actions and deeds of North Korea sympathizers. But nobody knows what to say if I ask them, "What do North Korea sympathizers do? I'm so curious to learn."
>
> I sometimes wonder if a few decades of collective efforts to build some autonomy into the community's daily lives [*jagi salm-ui ju-in-i doeneun*] was all in vain. Under the good years of lefty regimes, people with narrow

---

14 This incident brought huge media attention and social controversy after a woman dispatched to the island as a schoolteacher reported that she was a victim of a gang rape by male villagers and parents, especially because it was discovered that she was not the only such victim, as previous women teachers were silent and silenced about the sexual violence. See S. Hwang 2016; H. Jeong 2016.

> minds deeply influenced by ideological censorship from the time of the military regime hid their thoughts. Now they've become vocal and expose their base nature. It negatively affects community activities now, dampening the village atmosphere of decision making through dialogue and cooperation for the common good. It feels almost as if we are back living in the war or a refugee camp without any room to appreciate the value of mutual aid and collective action. Without knowing what's going to happen tomorrow, the only priority is to keep surviving in the short term. (Interview with the author, summer 2016)

Domin's narrative presents multiple layers of Pine Tree Hill's internal dynamics and its status as a post–Cold War core location.[15] First, Pine Tree Hill is scarred by the memory of refugees from the Korean War, and it has also relived Cold War censorship. We saw in the villagers' protest to the newspaper how the paper invoked the village's stigma as a war refugee town, one still poor to a point that the children playing online games were portrayed as the abandoned kids of broken families in a shantytown. The protest was important to villagers' sense of pride and ownership of their own history (*ju-in uisik*), so receiving an apology really mattered to them. The media protest reflected their dignity and their determination to shake off the stigma and shame of poverty during the post–Korean War and Cold War era. However, the political climate of the previous two regimes has enabled the interpellation of community activists as North Korea sympathizers, which renders community activism more difficult to separate and contest because the regimes posed protectionist positions by taking up populist demands for post–debt crisis economic recovery and social development.[16]

Since the division between North and South Korea has never overcome the status of an ongoing war, the Cold War presence offers the rationale of "national security" as the top priority for being subjected to US-led transpacific Cold War architecture in the name of alliances (Yoneyama 2016). During the Vietnam War, South Korea was the foremost ally of the US, functioning as a sub-imperial nation-state (Lee 2010) and accumulating national capital by supplying paid-labour soldiers and nurses

15 By the "post–Cold War regime," I refer to the disintegration of the former Soviet Union and Eastern European bloc, but I also use "post" here to align with postcolonial and poststructural theory and how those "posts" do not mean the cessation of colonial and structural entanglements (Shohat 2006; Yoneyama 2016).

16 The impeachment of Park Geun-Hye (10 March 2017) and the election of Moon Jae-In (10 May 2017) are other dramatic political changes whose consequences for daily politics and social relations need to be considered.

as well as accelerating industrial production, as Japan similarly aligned with the Americans and benefited from the Korean War (Glassman and Choi 2014). The US army still has more than fifty bases with approximately 690,000 personnel in the southern half of the Korean peninsula, amazingly segregated and hidden from mainstream domestic citizens' daily lives (Cheng 2010; Höhn and Moon 2010; Moon 1997; Yea 2015). However, during the recent democratic civilian presidential regimes, anyone who criticized state and municipal authority was called a communist in an attempt to isolate and nullify dissent. Although the interpellation process was regarded as a marker of the politically suppressed past, it has been revitalized in the public witch hunt of politicians and celebrities as North Korea sympathizers, along with the Park Geun-Hye regime's attempt to justify the past military authoritarian regime and its brutality by reference to North Korean threats.

Suspicion about leftists in association with North Korean sympathizers has been increasing since the 1990s – for example, over the so-called Sunshine Policy of the Kim Dae-Jung regime (1998–2003), which expanded trade and family reunions between North Koreans and South Koreans, and the Northern Limit Line dispute during the Roh Moo-Hyun regime (2004–8). However, suspicion was magnified in 2013, when the election fraud of the United Progress Party (*tongjin-dang*), an opposition party, exposed key members' likely involvement in advocates of the North Korea. It is no coincidence that 2013 was the beginning of the Park Geun-Hye regime. From 2013 to 2017, she not only articulated her right-wing ideological position most clearly, invoking her father, Park Chung-Hee, and his regime of 1961–79, but she also instrumentalized ideological suspicion as a political tool to suppress opposition parties and leftist politics (J-i. Kim 2014; Y-c. Kim 2014; Pak 2014).[17] The panic culminated in the "witch hunting" of leftists by conservative mass media, national assembly members, and juridical authority when they accused hosts of a public talk show of being North Korean sympathizers based on falsified claims, even though the hosts were attacked by the audience for an suicide attempt (Ahn 2012; S-j. Kim 2015; J. Jeong 2014; Yi 2015).[18]

By observing Pine Tree Hill community dynamics embroiled in haunting Cold War memories and their enunciations through local leaders, it is clear that community activism is not constantly homogeneous and

17 See Jeong's chapter in this volume regarding the background of Saemaul Undong's revitalization in relation to Park Chung-Hee's regime.

18 I thank Professor Kim Won for helping me trace the genealogies of the recent *jongbuk jwapa* (North Korean sympathizer) discourse. See also N. Lee 2007; Evans 2015; Song Ji-hye 2015.

unified. Pine Tree Hill is not merely contesting state authority as the administrator of welfare but is also struggling with the geohistorically produced architecture of Cold War regimes in the post–Cold War era.[19] It is impossible to rectify structural poverty when people's acts of sovereignty to change the social structure are interpreted as the work of communists aligned with North Korea.[20]

In this context of challenges stemming from the structural poverty that was triggered by and has been sustained since the Cold War, welfare became the predominant and omnipresent recipe to mediate the visible polarization of classes. Welfare is a hot potato issue in many countries, for both those with a long history of trying out different kinds of welfare states and those with a relatively short history of welfare that starts as neoliberal workfare or combines different models of welfare states (Kingfisher 2002; Smith 2007). In South Korea, during and after the Asian financial crisis, welfare was predominantly a politicized subject. It was used as a (neo)liberal governmental technology to mediate the crisis's socio-economic consequences, in addition to being a trope to mark South Korea's balance between social development (equated with democratization) and economic development (excelling in global market competition). As noted above, the Kim Dae-Jung regime that coincided with the Asian financial crisis launched the first welfare state, assuring everyone a basic standard of living premised upon universal welfare. Since then, welfare and the politics of well-being have been showcased by both the left and the right in elections – at presidential, provincial, municipal, and district levels – as the solution to social vice or class polarity that is aimed at appealing to voters. Welfare politics contributed to neoliberalization of everyday life in that it promotes discourses of certain affects and commodities of "enjoyment" and "well- being," whether they are food, vacations, or resort housing beyond domain of policymaking (Seo 2014; Jesook Song 2014; Zhang 2016; Žižek 2007). In other words, welfare politics and discourse govern a way of taking care of oneself in light of an overworked and precarious life.[21]

19 Although this chapter does not reveal the location of province, it is one of the provinces most heavily affected by McCarthyism in South Korea.

20 Eom's chapter in this volume also elucidates how this Cold War architecture and South Korean state sovereignty doubly marginalizes Chinese residents in South Korea. This marginalization is not only because of their being considered communists until recently but also, ironically, a result of viewing them as useful liaisons to the globally ascending People's Republic of China regime.

21 It would not be irrelevant to juxtapose this with Foucault's notion of technology of the self and his examples in Western history – for example, self-reflection in the ancient Greek period, confession during medieval times, administering in the modern epoch.

The Education Welfare Project as a Core Location

The Education Welfare Project in Pine Tree Hill can be viewed as a core location of distributional justice, primarily because of the ways in which it is in tension with the Children's Network. The EWP initatives are immanently "abortive" when social development is intended to counterbalance economic development: in other words, in this approach, social development exists only to ameliorate the consequence of economic expansion, and does not actually problematize the premise of development within capitalist systems. (I expand on the notion of "abortive" justice below; see also Trouillot 2000.). At the same time, these initiatives invite catachronic moments and unpredictable and disjointed possibilities (Yoneyama 2016) when subalterns try to be heard, rather than just seen, and actual changes are realized rather than remaining unsubstantiated symbolic gestures (Morris 2012).

When welfare was branded as the way for state and municipal governance to implement distributional justice as if it would resolve class contradictions and consequences of capitalist development, Pine Tree Hill residents and community activists opened up different ways to imagine welfare itself. Their direct actions and confrontations have negated social development's ventriloquilization of their efforts to build people's sovereignty. At other times, they have coasted together within and next to the operation of social development, despite the continued Cold War stigma against communism or criticism of universal welfare as financially inefficient. People's subalternity is not curated for populistic politics by privileging event-centred media politics or street protests as hegemonic modes of self-expression for the need of social transformation. Instead, true revolution is an outcome of parallel efforts in people's daily lives to transform mundane yet oppressive social relations (Morris 2012).

As anthropology theorists (e.g., Povinelli 2001 and Morris 2012) and Asia-as-method thinkers (e.g., Chen 2010 and Baik 2012) suggest, a place's geohistorical singularity allows for something new and even radical to emerge socially when its problems and ideas (*sasang*) are taken up by other thinkers as a way to imagine connections and forms of praxis. By engaging with the ideas behind Asia as method, the concept of core location can allow us to corroborate this radical potential in pursuit of a synchronicity of insights and the problematics against the hegemonic epistemology of modern sciences that seeks generalizability through reproducible and comparable capacities. Those ideas illuminate how the knowledge production of a location can contribute to decolonization and anti-assimilation when knowledge producers are embedded in the location's historical marginalities, especially when the very idea of core location is confronted by problematics created through empirically grounded research.

Welfare as Abortive Justice

Abortive justice is a notion I develop from Trouillot's concept of an "abortive ritual" of collective apologies (2000) and Yoneyama's discussion of post–Cold War redress culture in transitional justice (2016).[22] Trouillot's piece describes how increasing occasions of collective apologies are abortive rituals. His transatlantic context of collective apologies includes European Christians' repent of Crusade massacres and also President Bill Clinton's apology for American slavery to the ghost audience on the global stage. Trouillot clarifies that his question is not whether such apologies have any short-term benefits; rather, he seeks to reveal the conditions that make the wave of collective apologies possible. He problematizes the ways in which such apologies occur in particular illocutionary events that stabilize collective entities (both the apologizers and the addressed) in temporal and spatial distances, so that historical violence is addressed peacefully. In some cases, these apologies happened centuries after the atrocities were carried out, so the words of apology serve merely a performative function. Further, in that illocutionary structure that equalizes the positions between aggressors and victims, apologies of the former Western empire necessitate the Other, non-Western former empire subjects, to be in a forgiving position through a rhetoric of sharing pain, which obscures relations of power. He targets liberal juridical regimes as the crucial condition of making the wave of collective apologies possible by electing a particular mode of communication as hegemonic in the illocutionary events of collective apologies, which is inevitably abortive and infelicitous. The ways in which liberal ideologies elevate status or even empower the victims by making them "individuals" or "individual cultures" equal to aggressors or aggressor collectivity in the illocutionary position of saying yes or no to apologies (which are the only options) do not leave room for non-liberal modes of communication for redress to precede this ritual in ways that would be felicitous for the victims.[23]

---

22 "Transitional justice" refers to juridical reparation of inter-states' war crimes or the kinds of injustice that cannot be addressed in regular courts.

23 Christopher Krupa's (2013) discussion of the Truth Commission in Ecuador and Latin America resonates with the way in which redress is claimed only in the premise of disavowal of on-going violence.

Yoneyama also points out the limitations of the liberal juridical framework by problematizing transitional justice and redress culture in the transpacific post–Cold War regimes that are still embedded within the Cold War architecture. She focuses on post-1990s redress culture, particularly through the ways in which comfort women's issues have been politicized and addressed in state-to-state treaties along with tribunal courts of war crimes. She argues that the liberal juridical framework in transitional justice depoliticizes conflicts between nations or national subjects by seeking harmony and resulting normalcy of state sovereignty and global order inherited from the Cold War. One of the examples she provides includes the Asian women's fund that was established to compensate the victims on behalf of the Japanese government, which was refused by the majority of former comfort women. Also, her example of a Japanese ruling party leader's official apology regarding the atrocities endured by comfort women situates the double sides of redress culture: the official apology was exalted as noble because it was courageously performed despite the strong opposition within conservative party and nationalist protest to acknowledge the crime. At the same time, the apology for a crime against humanity wound up silencing a colonial legacy laden with gendered violence as well as any accountability for the Cold War security order in the Pacific. According to Yoneyama, this kind of rhetoric of crime against humanity is a tendency of liberal governance of transitional justice. That is, it is a form of transitional justice that deals with fundamentally destabilizing elements, such as the tension between a former colonizer and the colonized, without disturbing the structure by using a flattening universalizing discourse. In the context of comfort women, who were mobilized as sexual labour under colonial and imperial domination, the moment of redress of the violence bypassed US military hegemony in the transpacific that pacified the Japanese Empire while at the same time legitimized Japan's ascendancy in the region. Japan's official apology to Korean comfort women, which acknowledged that violence against gendered bodies represented a crime against humanity, is not simply novel, compared with the conservative nationalist voice in Japan. The official apology elevates comfort women's status as equal to that of humanity more generally, yet at the same time silences Japanese colonialism, in that Korean (and other ethnic) comfort women have never been equal to Japanese middle-class women, during either the empire or the US military occupation. Here, Yoneyama points out that epistemic violence occurs in the universalistic assumption of the moral economy of apology and forgiveness and in the further impossibility of real redress through transitional justice when harmonious humanities and equal grounds between victims and aggressors are imposed in exchange for legitimizing the sovereign state and normalcy of US hegemony in the relationship between Japan and Korea and the transpacific domain.

Both Trouillot and Yoneyama problematize the liberal juridico-political framework as a condition to make the abortive transitional justice possible in that it regards justice as setting up the victim and the aggressor as equals. If Trouillot's and Yoneyama's criticism of transitional justice is true – that it inherently prevents real redress – how possible is distributional justice in modern nation states, including welfare states? The need for redress emerges to recuperate former colonizers' or draconian states' moral legitimacy even when it relies on the rhetoric of liberating the oppressed, or securing human rights. In the context of the welfare state, I argue that it, too, is an *abortive ritual*. In general, welfare is politicized under the disguise of a utopian imagery or teleological view of democratic advancement in reference to the West. It equates the nation state form with being "advanced," so it graduates to the status of economically developed former colony, rather than grappling with the historically caused poverty and class polarization characteristic of this process.

I do not use the word "abortive" to mean miscarriage, as if the outcome is certain and its normalcy is predetermined. Following Trouillot, I choose "abortive" as opposed to "thwarted" or "failed" or "unproductive," because an unfulfilled reparation is not about a hampered process or no gain whatsoever. Instead, I am interested in a premise and framework that is not directed at securing reparations for wronged people. If the redress is premised on the forgetfulness of deeper violence or distraction through rhetoric of liberation or hopefulness, is it really different in the context of welfare?

At first, it might seem puzzling to think of the welfare state in the same domain as truth and reconciliation commissions or tribunal courts. The acts of a welfare state with respect to distributional justice are very much political economic matters that concern the subsistence levels of everyday people, whereas transitional justice is a political domain that hinges on the moral economy of apology and forgiveness through illocutionary singular events. After all, distributional justice mediates social relations that directly contribute to the capital accumulation process (or let us say national wealth), such as workers and populations that are essential to the reproduction of the workers in the name of dependants. Distribution is not a matter of moral economy but an essential component of understanding capitalist political economy, especially the necessary role of the state to mediate the labour and class relationship for capital accumulation. Thus, distribution is just as central as production to political economy, and, more importantly, distribution cannot be singled out and separated from production in the circuit of capital.

I consider the emergence of the welfare state as an abortive ritual of justice in capitalist modernity in dealing with the vulnerable and harmed subjects of historical violence in the process of capitalist accumulation. The social development and citizenship that arise in response to and buttress the capitalist accumulation process are not exceptional in geopolitical contexts where development is heavily embedded in Cold War security politics, such as South Korea (Glassman and Choi 2014; Lee 2010). In particular, South Korean history shows that the ascendency of the welfare state occurred during the Asian financial crisis in the course of the nation's self-criticism for the single-minded emphasis on economic development, where social development was meant to be a form of compensation. The ascendency of welfare in South Korea overlaps with proliferating reconciliation movements aiming to indemnify the historical violence of colonialism and transpacific Cold War crimes (Yoneyama 2016). This aspect of structural violence, whether stemming from South Korea's Cold War regime or its capitalist development, is recognized by the state and its attempt to redress it draws a parallel between welfare and reconciliation as abortive justice within liberal governing.

Conclusion

If the welfare state's acts of distributional justice are just as abortive as those seeking transitional justice under the similar premise and history of the liberalization of the justice domain that simultaneously opens up and erases reparation of historical and epistemological violence, what is at stake in pointing out this resemblance? Since my research on this question is still in an incipient stage of contemplation, I can offer only a tentative position. Nevertheless, an obvious implication of the juxtaposition between distributional justice and transitional justice lies in the material ground that welfare has become a frontier of solutions for neoliberal capitalism, whether being nostalgic about the Keynesian system in the European and North American contexts or the utopianization of welfare as social development to offset some of the effects of economic development in other part of the planet, as in South Korean context, or basic income movement in global scale.

A rather discrete implication is not unrelated to this obvious one, yet it is still in need of greater elucidation. I am concerned about how politically committed scholars, including anthropologists, who advocate distributional justice as the only practical option for dealing with the political economy of capitalism wind up making Marxism (although there are so many kinds) a culprit in ineffective counter-movements for dealing with capitalism. A typical criticism directed at Marxism is that Marxists tend to preoccupy themselves with problems of the accumulation process, rather than the distribution of the resources or consumption. For example, James Ferguson's efforts to address the significance of distribution in the form of basic income is understandable, yet unsatisfactory. Ferguson's position is understandable since he argues that, as the proverb goes, to give people fish instead of teaching them how to fish can ameliorate matters of immediacy, such as hunger; more importantly, he suggests a basic income model as an alternative to the Western welfare system by challenging the liberal premise inherent to welfarism that privileges independent individual as the deserving subject. This challenge to liberalism echoes Trouillot's and Yoneyama's criticisms of liberalism in transitional justice. Yet Ferguson's position is unsatisfactory because his suggestion of a basic income is not sufficient for minimal subsistence in the majority of cases, as he is acutely aware of. Further, the basic income model is built upon a false dichotomy between production and distribution as a solution to the problems surfacing from the capitalist system as if it is an option for not dealing with capitalist accumulation continuously. This illusive solution through a basic income model or distribution over a reckoning with the dynamics of accumulation resembles the paradigm of liberal duplicity in transitional justice in that it opens up a possibility of redress, but at the same time closes the door to the opportunity for or orientation of a fundamental reshaping of social relations. Here you see my concern with pragmatic assertions through distributional justice that render Marxism and political economy as ineffective and outdated approaches. The direction that I am heading with this observation is to suggest that a politics of distribution is an appropriation of political economy into a politics of moral economy by regurgitating a liberal framework that impedes the necessary process of confronting the destabilizing elements of capitalism in the very moment of opening up possible changes in a reconciliatory manner.

Therefore, I put those modes of justice together in the hopes of creating a dialogue between what is considered as "political" (focusing on the criticism of liberalism or making liberalism as the primary object of knowledge) and what is viewed as "political economic" (focusing on the ways in which capital accumulation is structured in the circuits of the production and distribution of commodities and social relations being construed both as a result and the engine of the reproduction of the social totality, both by Marxists and their opponents). Distributional justice through the welfare state is infelicitous in not dissimilar ways to how transitional justice through international courts (or truth and reconciliation commissions) is inherently abortive. Both deflect the orientation of reckoning with deep-seated structural issues by focusing instead on addressing negative repercussions in the name of practicality. The (neo)colonial order that haunts us in postcolonial redress culture is silenced when pressure to rectify structural conditions ends in mere nominal recognition or the logistics of monetary compensation. How the capitalist form of wealth has been and will continue to be made is sidelined when consequential problems of accumulation are redirected as a matter of generosity and a redistribution of wealth. These soothing rituals of liberal ideology that advocate for the rights of the vulnerable are not equipped to transcend deep-seated political-economic violence, such as colonialism and Cold War regimes. I assert that the EWP represents a prime example of a core location of distributional justice. It is a core location in that the EWP exposes the double marginalities of distributional justice – that is, with reference to the marginality of the working-poor class within the capitalist system and marginalization of a fundamental challenge to capitalism by focusing on distribution without questioning the wealth-making structure. At the same time, the EWP allows us to think of these marginalities as potentially spearheading grounds characterized by tensions and cooperation with neighbourhood sovereignty, rather than as sites of perpetual victimhood. We should take seriously, then, the idea that the EWP's singular constituency of the Children's Network in Pine Tree Hill serves as a platform for seeking justice for class contradictions that are not reducible to the sort of distributional justice pursued by liberal capitalist states.

REFERENCES

Abelmann, Nancy, Jung-ah Choi, and So Jin Park, eds. 2012. *No Alternative? Experiments in South Korean Education*. Berkeley: University of California Press.

Adnan, Shapan. 2015. "Primitive Accumulation and the 'Transition to Capitalism' in Neoliberal India: Mechanisms, Resistance, and the Persistence of Self-Employed Labour." In *Indian Capitalism in Development*, edited by Barbara Harriss-White and Judith Heyer, 23–45. London: Routledge.

Ahn, Young-min. 2012. "Jongbuk Manyeosanyang-e Ppajin Dangsin Deurui Daehanmin-guk" [Your Republic of Korea Has Fallen into a North Korea Sympathizer Witch-hunt]. *Minjok* [Nation] 21: 70–3.

Baik, Young-seo. 2012. "Sinjayujuuisidae Hangmunui Somyeonggwa Sahoeinmunhak: Ssun-geowaui Daedam" [The Intellectual Call in Neoliberal Era and Social Humanities: Dialogue with Sun Ge]. *Dongbangjihak* [Eastern Knowledge] 159: 421–71.

Baik, Young-seo. 2013. *Haeksimhyeonjang-eseo Dong-asiareul Dasi Mutda: Gongsaengsahoereul Wihan Silcheon-gwaje* [Rethinking East Asia in Core Locations: A Task for a Co-Habitation]. Paju, KR: Changbi.

Castel, Robert. 2002. "Emergence and Transformations of Social Property." *Constellations* 9 (3): 318–34.

Chatterjee, Partha. 2011. *Lineages of Political Society: Studies in Postcolonial Democracy*. New York: Columbia University Press.

Chen, Kuan-Hsing. 2010. *Asia as Method: Toward Deimperialization*. Durham, NC: Duke University Press.

Cheng, Sealing. 2010. *On the Move for Love: Migrant Entertainers and the US Military in South Korea*. Philadelphia: University of Pennsylvania Press.

Choo, Hae Yeon. 2016. *Decentering Citizenship: Gender, Labor and Migrant Rights in South Korea*. Stanford, CA: Stanford University Press.

Donzelot, Jacques. 1984. *L'invention du social*. Paris: Vrin.

Donzelot, Jacques. 1988. "The Promotion of the Social." *Economy and Society* 17 (3): 395–427.

Evans, Steven. 2015. "Why South Korea Is Rewriting Its History Books." *BBC News*, 1 December. Accessed 20 July 2017. http://www.bbc.com/news/world-asia-34960878

Glassman, Jim, and Young-Jin Choi. 2014. "The *Chaebol* and the US Military-Industrial Complex: Cold War Geopolitical Economy and South Korean Industrialization." *Environment and Planning A* 46: 1160–80.

Höhn, Maria, and Seungsook Moon, eds. 2010. *Over There: Living with the US Military Empire from World War II to the Present*. Durham, NC: Duke University Press.

Hwang, Sun-min. 2016. "Jibdan Seongpokhaeng Imyeone Kkallin 'Pyeswaejeok Gongdongche'ui Jibdan Beomjoe" ["Exclusive Community"s Collective Crime behind Gang Sexual Violence]. *Maeil Business Newspaper* website, 5 June. Accessed 21 July 2017. http://news.mk.co.kr/newsRead.php?no=403603&year=2016

Hwang, Unseong. 2007. "Nosugin Bokji 10 Nyeon, Saeroun Jeonhwanui Mosaek" [10 Years of Homeless Welfare, Seeking a New Turning Point]. In *Seoul-si Nosugin Jeongchaek Toronhoe* [Seoul City Homeless Policy Debate], 8–33. Seoul: South Korean Anglican Church Vision Training Center.

Jang, Jin-ho. 2011. "Neoliberalism in South Korea: The Dynamics of Financialization." In *New Millennium South Korea: Neoliberal Capitalism and Transnational Movements*, edited by Jesook Song, 46–59. New York: Routledge.

Jeong, Hoe-seong. 2016. "Sinan Seommaeul Seongpokhaengbeom 3 Myeong 18~12 Nyeon Mugeoun Jing-yeokseon-go" [In Sinan Island Village 3 Individuals Charged with Sexual Assault Are Given a Heavy Sentence of 18~12 Years]. *Yonhap News* website, 13 October. Accessed 20 July 2017. http://www.yonhapnews.co.kr/bulletin/2016/10/13/0200000000AKR20161013118900054.HTML?input=1195m

Jeong, Jeong-hun. 2014. "Hyeomo Wa Gongpo Imyeonui Yongmang: Jongbuk Damnonui Silche" [Desire behind Hatred and Fear: Reality of North Korea Sympathizer Discourse]. *Urigyoyuk* [Our Pedagogy]: 96–103.

Joseph, Miranda. 2002. *Against the Romance of Community*. Minneapolis: University of Minnesota Press.

Kim, Eunjung. 2017. *Curative Violence: Rehabilitating Disability, Gender, and Sexuality in Modern Korea*. Durham, NC: Duke University Press.

Kim, Jeong-in. 2014. "Jongbukpeureimgwa Minjujuuiui Wigi" [The "Following-North" Frame in Ideological Conflicts and Crisis of Democracy]. *Yeoksawa Hyeonsil* [History and Reality] 93: 209–33.

Kim, Seo-jung. 2015. "Jongbukmori, Eollondo Hanmok Haetda" [Following-North, Mass Media Is a Contributor, Too]. *Hwanghaemunhwa* [Yellow Sea Culture] 86: 352–58.

Kim, Suhyun. 2001. *Seoul-si Homniseu Yeoseong Siltaewa Daechaek* [The Conditions and Measures to Address Homeless Women in Seoul City]. Seoul: Seoul Development Institute.

Kim, Yun-cheol. 2014. "Maidongpunggwa Jongbuk Mori: Bak Geun-hye Jeonggwonui Tongchi Jeollyak" [Chasing North Korea Sympathizers While Turning a Deaf Ear: The Dominant Strategy of Park, Geun Hye]. *Simingwasegye* [Citizen and the World] 24: 12–23.

Kingfisher, Catherine. 2002. *Western Welfare in Decline: Globalization and Women's Poverty*. Philadelphia: University of Pennsylvania Press.

Krupa, Christopher. 2013. "Neoliberal Reckoning: Ecuador's Truth Commission and the Mythopoetics of Political Violence." In *Neoliberalism, Interrupted: Social Change and Contested Governance in Contemporary Latin America*, edited by Mark Goodale and Nancy Postero, 169–94. Stanford, CA: Stanford University Press.

Lee, Jin-kyung. 2010. *Service Economies: Militarism, Sex Work and Migrant Labor in South Korea*. Minneapolis: University of Minnesota Press.

Lee, Kang-Kook. 2011. "Neoliberalism, the Financial Crisis, and Economic Restructuring in Korea." In *New Millennium South Korea: Neoliberal Capitalism and Transnational Movements*, edited by Jesook Song, 29–45. New York: Routledge.

Lee, Namhee. 2007. *The Making of Minjung: Democracy and the Politics of Representation in South Korea*. Ithaca, NY: Cornell University Press.

Moon, Katharine. 1997. *Sex among Allies: Military Prostitution in US-Korea Relations*. New York: Columbia University Press.

Morris, Rosalind, ed. 2012. *Can the Subaltern Speak? Reflections on the History of an Idea*. New York: Columbia University Press.

Pak, Kwon-il. 2014. "Jongbukmoriui Naejeok Nolli: Chiansahoeron" [Ulterior Motive of Painting Following North: Paradigm of Security Society]. *Hwanghaemunhwa* [Yellow Sea Culture] 82: 48–64.

Park, Hyun Ok. 2011. "For the Rights of 'Colonial Returnees': Korean Chinese, Decolonization, and Neoliberal Democracy in South Korea." In *New Millennium South Korea: Neoliberal Capitalism and Transnational Movements*, edited by Jesook Song, 115–29. New York: Routledge.

Park, So Jin, and Nancy Abelmann. 2004. "Class and Cosmopolitan Striving: Mothers' Management of English Education in South Korea." *Anthropological Quarterly* 77 (4): 645–72.

Povinelli, Elizabeth A. 2001. "Radical Words: The Anthropology of Incommensurability and Inconceivability." *Annual Review of Anthropology* 30: 319–34.

Sanyal, Kalyan. 2007. *Rethinking Capitalist Development: Primitive Accumulation, Governmentality and Post-Colonial Capitalism*. New York: Routledge.

Seo, Dongjin. 2011. "The Will to Self-Managing, the Will to Freedom: The Self-Managing Ethic and the Spirit of Flexible Capitalism in South Korea." In *New Millennium South Korea: Neoliberal Capitalism and Transnational Movements*, edited by Jesook Song, 84–100. New York: Routledge.

Seo, Dongjin. 2014. *Byeonjeungbeobui Natjam: Jeokdaewa Jeongchi* [Daydream of Dialectics: Oppositions and Politics]. Seoul: Currie.

Shin, Kwang-Yeong. 2011. "Globalization and Social Inequality in South Korea." In *New Millennium South Korea: Neoliberal Capitalism and Transnational Movements*, edited by Jesook Song, 11–28. New York: Routledge.

Shohat, Ella. 2006. *Taboo Memories, Diasporic Voices*. Durham, NC: Duke University Press.

Smith, Anna Marie. 2007. *Welfare Reform and Sexual Regulation*. Cambridge: Cambridge University Press.

Song, Jesook. 2009. *South Koreans in the Debt Crisis: The Creation of a Neoliberal Welfare Society*. Durham, NC: Duke University Press.

Song, Jesook. 2014. *Living on Your Own: Single Women, Rent Housing, Post-Revolutionary Affect in Contemporary South Korea*. Albany, NY: SUNY Press.

Song, Jesook. 2017. "Gendered Care Work as 'Free Labor' in State Employment: School Social Workers in the Education Welfare (Investment) Priority Project in South Korea." *Journal of Asian Studies* 76 (3): 751–68.

Song, Ji-hye. 2015. "Haewoe Hakjadeul 'Han-guk Jeongbu Gukjeonghwa, Abewa Ddokgatda'" [Scholars Abroad Declare "the Korean Government's Policy Implementations and Prime Minister Abe Are the Same"]. *Sisain* 425. Accessed 23 May 2019. https://www.sisain.co.kr/?mod=news&act=articleView&idxno=24657

Trouillot, Michel-Rolph. 2000. "Abortive Rituals: Historical Apologies in the Global Era." *Interventions* 2 (2): 171–86.

Yea, Sallie. 2015. *Trafficking Women in Korea: Filipina Migrant Entertainers*. New York: Routledge.

Yi, Hang-u. 2015. "Jongbukui Jeongchisahoehak" [Political Sociology of Following North]. *Han-guksahoehakhoe Sahoehakdaehoe Nonmunjib* [Korean Sociology Association Conference Proceedings]. December, 53–4.

Yoneyama, Lisa. 2016. *Cold War Ruins: Transpacific Critique of American Justice and Japanese War Crimes*. Durham, NC: Duke University Press.

Zhang, Li. 2016. "The Rise of Therapeutic Governing in Postsocialist China." *Medical Anthropology* 35 (2): 6–18.

Žižek, Slavoj. 2007. *For They Know Not What They Do: Enjoyment as a Political Factor*. London: Verso.

7 Situating the Space of Labour: Activism, Work, and Urban Regeneration

SEO YOUNG PARK

Since 2012, young scholars, artists, and activists have changed the social and physical landscape of Changsin-dong, a disenfranchised town populated by low-income residents and garment workers. This reconfiguration involved various social enterprises cooperating with the municipal government, working together to design a business model that aims to maximize its "social impact" by providing employment opportunities, supporting local communities, and helping the environment. Seoul has seen an increasing prevalence of this new form of civic participation in the public sector, local community, and urban governance, especially during the first two terms of Park Wonsoon, who was elected mayor of Seoul in 2011, 2014, and 2018. These attempts are often considered to have replaced the old model of state-led destructive urban developmentalism with a new paradigm of "regeneration" – that is, the mending and reusing of space (instead of destruction and reconstruction). A recent news article even subjected the concept of regeneration to homonymic word play: from Changsin (昌信, the official name of the town) to Changsin (創新, creating anew).[1]

The new logic of "regeneration" in Changsin-dong was notable as it shows the state's social and urban projects of re-branding a location and reframing the existing notion of developmentalism in South Korea – one of the running themes of this volume. It was a significant case for many, as the neighbourhood had long been excluded from the changes and development of other parts of the city. While the adjacent Dongdaemun

1 Eunhwa Han, "Jaejomyeong Bannen Seoul Jongnogu Changsindong" [The Re-evaluation of Changsin-dong]," *Jungang Ilbo*, 15 May 2015, accessed 26 May 2019, https://news.joins.com/article/17813420

Market[2] transformed from a manufacturing base for global fashion and the garment industry (from the 1960s to the 1980s) into a hub of the transnational fast-fashion shopping (after the late 1990s), Changsin-dong, just one street across from the market, has been populated by former and current garment-manufacturing businesses. Changsin-dong's small-scale home-factories are based on labour-intensive systems and are a residual space that has served, yet was left behind by, the fast industrialization and transformation of Seoul. The recent new urban branding has attempted to actively promote the presence of small-scale workshops by beautifying the dark allies where shanty workshops were tucked into residential buildings, enhancing the liveability for residents, and making the town more inviting for visitors and tourists. The word *hyeonjang* – a scene, field, site, or location in Korean – has been prominent in descriptions of this "regeneration" project of this neighbourhood. The town has been branded as a place where people are able to feel *nodong-ui hyeonjang* (a labour scene or field) or, more specifically, *bongje nodong-ui hyeonjang* (a garment stitching labour scene or field), which is otherwise not readily visible or accessible in the heart of the rapidly transforming mega-city of Seoul.

Hyeonjang was also a term that I frequently heard from former garment workers and activists when I was conducting an extensive part of my fieldwork in the adjacent Dongdaemun Market from 2008 to 2010 (Park forthcoming). For former activists, the term "field" referred to a site of praxis where intellectuals and activists can conceptualize a problem and materialize their political ideals in dealing with the repression of the South Korean military state and capitalist corporatism under the auspices of the US–Japan transpacific alliance. For garment workers, it referred to the site of their everyday work lives. Quite a few garment workers I interviewed have worked in Changsin-dong and its vicinity for decades and had actively participated in union activity in the 1960s and 1970s. During the 2000s, entrepreneurs, urban developers, and community activists began to pay attention to the ongoing social network of garment manufacturing in Changsin-dong; and the neighbourhood emerged as a site full of historical memories and new forms of social experimentation.

By tracing the varying meanings of the word *hyeonjang* ("field") that emerged in Changsin-dong and by exploring the way they impact one another, this chapter unravels the multiple layers of marginality

---

2 The marketplace was historically formed around the Dongdaemun (Grand East Gate) of the city of Seoul during the Joseon dynasty (1392–1897).

of this neighbourhood as a space of garment labour. This physical, social, and epistemological space of "field," in turn, resonates with the ethnographic field site in which I as a researcher was situated. While Changsin-dong is usually remembered as a town of industrial ruin and marginalization, its genealogies of garment labour and labour activism invoke the possibility of dynamic change and have brought it to the forefront of new urban experiments. For this analysis, I first draw on my previous ethnographic research, where I worked with garment workers, residents, and former labour activists in the Changsin-dong area, to contrast the two different meanings of "the field" by former activists and garment workers. I then explore the shifting interests in the "field of garment labour" around the new urban project. Finally, I analyse the changing notions of "field" within a discussion of ethnographic field sites and core locations (*haeksim hyeonjang*) (Baik 2013, 47). Through this, I argue that the notion of "field" is construed in multiple temporal orientations – in its disclosure and persistence, stagnation and revitalization, and nostalgia and speculation about the future – and challenges the dominant teleological narrative that has framed the paths of capitalism, activism, and fieldwork alike. In so doing, I reconsider the concept of "core location" (*haeksim hyeonjang*) as an analytic apparatus to approach empirical studies and engage attentively with the complex implications of *hyeonjang* as a lived space.

The Disappearing Field of Praxis, the Continued Space of Work

During my fieldwork in 2008–10 in the Dongdaemun area, including Changsin-dong, I encountered numerous women who had been formally involved in garment work and labour activism. With a particular interest in the politics and temporality of the city and work experiences, I paid close attention to the changing conditions of labour and, more importantly, to subjective narratives of work throughout people's lifetime. The subjective narratives often made different articulations regarding their long experiences working in the garment industry and their perspectives on the current situation. For instance, Munhee, a formal member of the Cheonggye garment union, talked about the helplessness she felt about "the field": "Things have changed so much and the *field has almost gone* in a sense. They [the workers] are just all dispersed and it is impossible to make a change or mobilize some energy" (interview with the author, 15 January 2010; my emphasis). Munhee was referring primarily to the transition of garment manufacturing in urban areas from a centralized, mass-manufacturing process to outsourcing commodity chains dispersed to other parts of Korea and overseas to take advantage of lower labour costs. At the same time, Munhee was also suggesting the impossibility of continuing the spirit of collective union work with the "bygone" field. In contrast, I also encountered different accounts of current stitching workers, who described the field of work as rather persistent and continuous. While the changing conditions of employment and their implications for labour rights has been observed in a broader context in general, "the field" as a lived space is situated in the personal, social, and discursive histories of both activists and workers, as I explore in this section.

Since the 1960s, the neighbourhood of Changsin-dong has been formed in the midst of the evolution of the Dongdaemun garment market, located across the street. Behind the current facade of Dongdaemun's shopping and tourist scene, as a massive cluster of garment wholesale and retail shopping plazas and the iconic DDP (Dongdaemun Design Plaza), the old structures of former warehouses and factories indicate its prior history of labour-intensive manufacturing from the 1960s to the 1980s. During the heyday of mass production, Dongdaemun's buildings were filled with factories on the upper floors and wholesale shops on the ground level. The authoritarian government implemented an export-centred economic developmental plan, fuelled by a low-grain price policy based on US food "aid" that allowed for low-cost wage labour (Cheng 1990),[3] which included labour-intensive garment manufacturing as one of the central industrial sectors. This broad economic plan forced a large number of underemployed rural people into the city to seek jobs (S.k. Kim 1997; Koo 2000). If a city like Pohang was a central site of "heavy industry" (see Jeong's chapter in this volume), Dongdaemun Market engaged in "light-industry," providing mass-manufactured clothes to a growing domestic ready-made clothing market as well as for foreign apparel corporations. Young female workers were mobilized as cheap labour within the gendered and patriotic narrative of "industrial warriors." Multiple demands were made on women's bodies while their work and workplaces were marginalized (see, for example, Bonacich and Appelbaum 2000; Mills 2003; S.K. Cho 1985; S. Chun 2017).

Changsin-dong was regarded as a symbolic space for marginalized garment labour, as the neighbourhood housed workers who walked to their factories across the street, if not sleeping in the dorms on the top

3 See also Oh's chapter in this volume for the relationship between state planning and uneven development.

floor of their factories. It was also the neighbourhood of the legendary figure Chun Tae-il, a fabric cutter and the prominent labour activist of Cheonggye garment factories. As the conditions of garment labour became harsher, the garment labourers sparked vehement activism in the form of labour unions that represented the front line of democratic movements against the military government. The garment movements in the area impacted many other subsequent forms of activism, such as the anti-poverty movement led by the figures discussed in Cho's chapter in this volume. The alliance that was formed among workers and young intellectuals holds a significant place in the contemporary history of South Korean activism for labour rights, political liberty, and social equality and justice.

During this period, the widely shared idea of "field," or *hyeonjang*, meant a site or a base where the praxis of knowledge, labour, and transformative action could take place simultaneously. Left-leaning Korean activists, particularly college students and those with backgrounds in higher education, pursued their interest in a coalition with blue-collar workers in the manufacturing sectors (J.J. Chun 2011, 79). The idea of "field," especially for those who focused on mass-based class struggles, emphasized that political awareness and subjectification should happen in the everyday workplace along with their coalition with other labour subjects and activists. Numerous formal labour activists, students, and researchers were known to have infiltrated workplaces through *wijangchwieop* (getting a job with their real identity and motives concealed) in order to investigate labour conditions and collaborate with labourers (H.Y. Cho 1988; Yoo 2013). However, the suppression of unionism under Chun Doo-Hwan's military regime (1980–88), based on the National Security Law, caused numerous unions to disband. The changing regime of labour in the late 1990s triggered an "organisational crisis" (J.J. Chun 2011, 74), within unionism and collective action based on its foundation of regular employment, a shared identity among participants, and a centralized structure for the movement.

Moreover, the 1990s saw a shift away from Fordist, mass-production methods in the garment industry and a neoliberal transition of labour deregulation, impacting both organized unionism and the nature of garment manufacturing. Seeking lower labour costs, mass-scale outsourcing factories (of both Korean and foreign apparel companies) began to move to overseas cities such as Guangzhou or Hanoi in the early 1990s. Garment production, accordingly, has faded away in Seoul, leaving only small- to mid-scale workshops dispersed throughout the city. At the same time, the emerging need of just-in-time production turned the main function of Dongdaemun Market to a retail site for local and global consumers, taking advantage of the remaining and experienced workers, without ameliorating the conditions of their labour.

It is within this context that some formal activists with whom I conversed considered that their "field is gone." Changsin-dong and its neighbouring Dongdaemun and Gangbuk districts have gradually become a contested site under the changing conditions of post-Fordist production and the decline of unionism. At the beginning of the rise of labour unions in the early 1970s, labour standard laws were emphasized and the mass-scale factories closed after business hours. Some factories moved to Changsin-dong by downsizing their business, and manufacturing workers from the mass-production site started small workshops in their own houses or converted residential spaces to small scale-workshops that I call "home-factories" (S.Y. Park 2012). Gradually, these small units operated more effectively for the quick in-season turnaround of the Dongdaemun Market, yet they turned out to be on the "margin" and thus not conducive to traditional forms of unionism. On this, Munhee stated, "They do things in such a family way and are personally connected. Workers are paid by the piece so they are only interested in getting more orders and making more [clothes]. The industry is gone too anyways, as cheap clothes from China flood into [Korea]. They all have different issues and there is not much we can organize collectively" (interview with the author, 15 January 2010). The "family way," as Munhee referred to it, indicates the "informality" of workplace relations and practices (*gajok kkili jumeokguguro*). While the small-scale, decentred, and informal subcontract relations and manufacturing style formed a favourable condition for the "quick" adjustment in flexible garment production, this condition prevented both the standardized and competitive production of commodities and the possibility of mobilizing for formal labour advocacy. As Munhee returned to the garment factories in the 1990s after unions became normalized again, she found this situation helpless. During the time of my fieldwork, factories in Changsin-dong and its adjacent neighbourhoods were considered residual and endangered sites that no longer held any significance in garment manufacturing from a macro-scale point of view, for its focus on producing cheaper, non-branded clothes or producing at smaller scales.[4]

4 For larger-scale garment manufacturing made for the top- or second-tier garment market, such as designer boutiques or well-known brands, factories tend to be located in the eastern or southern part of Seoul.

The cascade of "cheap Chinese products" mentioned in Munhee's comment was often identified as the source of the crisis in my frequent encounters with various actors, including policymakers, labour activists, and business owners. The global reconstruction of garment manufacturing compelled developing countries such as South Korea to compete within the labour market. The labour cost theory determined the economic positions of Third World labourers and then arranged them into a comparable hierarchical ladder. Likewise, there was a prevalent perspective that saw the garment production in overseas factories and the immigrant labourers in local factories as an "inevitable threat" to the South Korean economy and to Korean labour, while the aging of the garment worker population has intensified over the past two decades. Most of the trained Korean stitching workers were over forty, as many interlocutors complained that "today's Korean young people would seldom consider working in the garment factory." As one of my interlocutors jokingly commented to me, "You can no longer do undercover research like others [in the past],"[5] because a young Korean woman in her thirties would stand out in today's factories. There might be one or two people under forty in a factory, and most of them are Chinese Korean, Nepali, or Vietnamese. Many of the new stitching assistants were often not able to sustain their jobs for long, due to the unstable employment conditions. Moreover, frequent police crackdowns, due to Changsin-dong's relatively close location to the centre of Seoul, discouraged numerous factories from hiring migrant workers. The aging of its residents and workers suggests not only a demographic change, but it was also experienced as a lack of the reproduction of labour and the loss of political spirit that had once been prominent in this field.

However, contrary to the former union activists' sense of their field having come to a close, current workers' narratives interpreted their field site as being more prolonged and continuous. Jisun, a sewing worker in her mid-fifties, remembered the changes in the mode of production and described how she has felt about her own "field" of work: "If you keep working *in the field* [*hyeonjang*], you will see people leaving the work a lot but also coming back to work as well ... I have been doing this work for too long. Life has been always harsh and it is nothing new ... I know there were reporters and students like you that also came in at different times and then left. [But] It does not change what

5 Seung-kyung Kim reflected upon the employment conditions in the late 1980s and the challenges of conducting participant observation (S.k. Kim 1997, x–xi).

I do" (interview with the author, 18 March 2009; my emphasis). Living and working in Changsin-dong for more than thirty years, and once a participant in the labour movement of the 1980s, Jisun sympathized with what former union leaders said and acknowledged how the dispersed workshops did not seem the same. Many of the garment workers that I worked with had vivid memories of attending night classes (*yahak*) in Christian churches after work, which were led by college students and older activists. Some of them were involved in garment labour union activities and participated in the militant and fervent protests that took place in Dongdaemun and other manufacture sites. Even those who did not directly join the union activities had varying degrees of contact with and memories of the democracy and labour movements, and often communicated with their fellow colleagues who actively joined the protests.

Yet, unlike many of the union activists, many garment workers persistently worked in the occupation, and this "field" of work for them has always been embedded in their everyday lives. Even though the conditions of the work and activism have changed, their everyday lives running the sewing machine to ensure an income for their families have not changed, from the time of union work to the present. The field of work, in this sense, could not have simply vanished, regardless of the scale. For the workers, the flexibilization of production and the changing frames of their work stressed resilience and persistence of their everyday practices. Workers' personal connections and subjective interests intersect with the intricate outsourcing factories and workshops that constitute the supply chain in Changsin-dong. Subjective narratives of workers and their genealogies of garment skills, obtained during my fieldwork, made me view the site of small-scale stitching work not merely as a vanishing form of informal economy, but as an intimately intertwined site of life, social relationships, and manual skills (S.Y. Park 2012).[6] The "field" of labour for my interlocutors has been entangled with disparate momentums and historical contingencies, as their life course dedicated to the garment industry has lasted much longer than the prime time of labour activism and a period marked by a particular mode of production in the area.

---

6 Recent ethnographies call for attentiveness to the specific practices and skills in manufacturing sites that are not fully explained by the macro-perspectives of garment labour. For the case of China, see Nellie Chu's work on Guanzhou's garment district (2016).

"Reporters and students like you," as Jisun mentioned in the interview quoted earlier, highlights the temporariness of external gaze on garment workers. While the major focus of labour activism shifted to larger-scale manufacturing sites and major issues such as migrant labour or irregular employment in the late 1990s and early 2000s, other interested actors and media sporadically came to the neighbourhood and conceptualized garment workers from different angles. In a group interview with Jisun and her friends, they talked about the different terms used to portray garment workers in Changsin-dong. In some of the documentaries they were portrayed as the "protagonists" of industrial times (*saneophwa sidaeui juyeokdeul*); while in a popular television show, they were "the masters of life" (*saenghwarui darin*). Jisun said there are "celebrities" in the neighbourhood who became popular by being featured on TV from time to time. There were also new social activitsts who approached Changsin-dong with various forms of expertise – social welfare, childcare and education, or social enterprise. These activists sometimes framed the workers as significant economic actors of garment manufacturing or as potential "artisans" (*jang-in*) with high-quality skills.

Although with varied interests and small in number, these new activists kept bringing new dynamics to the neighbourhood and maintained its symbolic value as a "field of garment labour" visible to the wider Korean society. Their projects in Changsin-dong extensively influenced my fieldwork, as they provided me with differing access to garment workers and also a perspective from which to analyse their labour politics in the context of the early 2000s. Workers such as Jisu often asked if I read articles in which Chansin-dong workers were featured or if I knew particular names of other researchers or students who have come to work with them. During my interviews, workers often reflected on these interactions to explain how they made them think about their own work. The collaborations of activists, residents, and workers sometimes remained as a continuous community or casual social networks with various levels of engagement, which made an impact on the way Changsin-dong became an exemplary case for the new Urban Regeneration Program that I explore in the following section.

The Shadow of Urban Developmentalism, the Forefront of Urban Regeneration

After I finished the extensive period of my fieldwork and left Seoul, the history of Changsin-dong's presence as "a field of labour" has gained a new meaning through an urban regeneration project that began around

2012. Ironically, Changsin-dong's long-term marginalization and the fame of its garment factories attracted social enterprises, community builders, and the metropolitan government to it as a prominent site of new urban experiments.

In many cases, urban housing policy in South Korea has been characterized by the complicit collaboration of large-scale business conglomerates and the authoritarian state to bring about top-down, coercive, and massive construction projects for rapid urban growth from the 1970s to the 1990s (see Hae's discussion on the construction state [*togeon-gukga*] in this volume). Since its enactment in 2002, led by then-mayor Lee Myung-Bak, the "New Town Program" claimed its awareness of the class and regional disparity between the north and south sides of the Han River, which has been regarded as one of the critical problems spawned by earlier development plans. However, the program revealed itself to be not much different from the previous growth-centred model (Shin and Kim 2016), given its focus on the real estate–centred program of the previously underdeveloped area that led to a rise in real estate prices before the actual renewal process was completed (Byeon 2013).

Changsin-dong, along with its adjacent Soongin-dong area, provides prime examples of the contradictory results of the large-scale, state-led New Town Program (Shin 2017, 549–50). When the two districts were assigned to the New Town Program in 2007, they drew attention due to their being the largest project areas (846,100 m^2) with a highly dense, marginalized population of 24,524. However, the renewal process did not achieve much progress, since the program did not account for a number of the specificities of the targeted areas. First, the two areas were contiguous with the national treasures of Dongdaemun (Grand East Gate) and the restored Seoul Wall. The city of Seoul has been trying to list these sites with the UNESCO World Heritage Centre, which perforce meant construction was restricted in the area. Second, many residential houses were ambiguously owned and occupied by unregistered residents from the early post–Korean War period. Third, including stitching home-factories and garment supply stores, there are numerous mixed-use buildings, used for both residential and commercial purposes. During my fieldwork, I frequently heard that my interlocutors, who mostly worked as part of these small-scale supply-chain units, were concerned about losing their base for home-factories under New Town Program, as their proximity to the Dongdaemun garment market was critical for their survival. A significant number of the residents opposed the New Town Program, which planned a collective reorganization of the uses of the buildings to replace them with high-rise apartment complexes. By 2012, 85 per cent of the proposed New Town developments had not broken ground, and the new mayor, Park Wonsoon, made a public apology for the lack of progress (Berg 2012), which ultimately signalled an official acknowledgement of the program's failure.

In 2013, the Changsin-dong and Soongin-dong areas became the first cases in which residents voted to be removed from the New Town Program, followed by many others, and thereby later became pioneering cases for Park's alternative Urban Regeneration Program. As Mayor Park strategically abandoned the New Town Program, the Urban Regeneration Program emphasized a placed-based and locally specific revitalization plan that emphasized public funding and resident participation.[7] In contrast to the standardized top-down policy that ignored local differences, the program promoted its dedication to the "cultural identities and memories" and actual needs of the local residents. For this purpose, Changsin-dong's primary local identity was that the neighbourhood was one of the rare "scenes" (*hyeonjang*) for economic practices based on locational specificities – that is, many residents were actually conducting their livelihoods within the neighbourhood, through various registers of garment making.

The program relied heavily on individual groups that could immerse themselves in and animate this local livelihood, combining place-based activism and the new regime of social enterprise in South Korea. Ranging from a research and business unit connecting urban-planning researchers with garment factories, to fashion designers, to public artists, these groups collaborate with each other, extend to different areas and networks with similar foci, and reconnect with former and existing individual activists and organizations in Changsin-dong. At the same time, the groups and organizations draw on diverse funding sources, which include large-scale corporate entities such as the Hyundai Motor Company and the Korean Mecenat Association (a corporate-funded non-profit organization that channels corporate enterprises' contributions to art and cultural activities); public funds such as Mayor Park Wonsoon's Hope Fund (*Huimang Peondeu*); micro-finance entities such as the Social Solidarity Bank (*Sahoeyeondae Eunhaeng*); and state-sponsored grants from district and central governments such as the Changsin-Soongin Urban Regeneration Promotion Center (*Dosijaesaeng Jinheungsenteo*). These financial and institutional networks formed the

7 While I focused on the specific policy of Seoul, in the 2000s there have been various initiatives by state and local governments to bring about urban regeneration, aiming to enhance both liveability and social welfare. Song's chapter in this volume deals with similar forms of government support in a different urban location.

basis for "social" programs driven by a combination of South Korea's state-led economic development trajectory and a globally emerging neoliberal regime of social welfare provision (M.Y. Cho and Lee 2017). These programs evolved in Changsin-dong from a public intiative to create employment for local garment workers in social enterprises (Park forthcoming). Over time, the total number of registered social enterprises radically increased, from 55 in 2007 to 1,526 in 2016 (Korea Social Enterprise Promotions Agency 2016), and Changsin-dong has become a leading target for these enterprises.

Ultimately, the Urban Regeneration Program involved diverse social activists, artists, scholars, corporate capital, and the metropolitan government, all of whom claim to have an engaged relationship with the residents, revealing the dynamic forces that have formed the space of inherent marginality. The enduring marginal status of the residents, the presence of labour and social activists, and the "underdeveloped" neighbourhood that has not yet been overtaken by large apartment complexes made the area an ideal site for a project aimed at "regeneration." The Urban Regeneration Program in Changsin-dong engaged with the "marginal population" by promoting the formally ignored and sacrificed garment workers and converted their space of work into a place-marketing strategy.

The "relevance" that the neighbourhood garnered, as a "frontier" of the Urban Regeneration Program, is unfolding in two ways, complicating the politics of change it would bring. On the one hand, the newly engaging social enterprises extended the scope of social relations that can be formed around garment work. As a recent ethnography conducted by Han (2017) has revealed, the young leaders of these projects come from various backgrounds, including those who grew up in the neighbourhood, those who have been deeply engaged for a long time with the childcare support program in the community, and those who were attracted to the neighbourhood as a site of social experimentation (Han 2017, 83–92). A prominent social enterprise redefined the neighbourhood as a "town of makers" – those who make something through their manual work, including garment-stitching labour. These examples represent efforts to create a new relevance between the work and the local community that did not exist or was not visible before. In a similar manner, these organizations connect local residents, garment workers, and activists with practices of art, education, and radio broadcasting, and different kinds of market networks from the supply chain of fast-fashion manufacturing. The space of garment labour, then, is presented and imagined as a web of multiple potential connections, rather than as a contained and isolated form of subcontracted work.

On the other hand, this manufacturing "field" was rendered an object of aesthetic, visual consumption for outsiders as a symbol of the "past" – what Cho calls "the narrative of past" in her chapter in this volume. It is not just the landscape but also the actual process of labour that has become fetishized. In preserving and reusing the old, decaying forms of old houses, convoluted narrow allies, fabric leftovers, and the outdated marketplace and representing them in an edgy and retro-fashion style, it has enhanced the outlook of the street and imbued the desolate neighbourhood with a youthful and inviting atmosphere for visitors. The beautification of the built environment has been the most prominent feature of the recent changes, as they focus on place-based art and architecture projects to enhance the liveability of the physical environment. The Urban Regeneration Program has stiven to discover the historical and social "content" of the neighbourhood to connect with retail, food, and lodging to trigger economic revitalization (Seoul Metropolitan Government 2016). The project became widely known for its active participation in the *Made in Changsin-dong* exhibition at the Seoul City Museum in 2013 and the "urban stroller" program that the museum designed in conjunction with this exhibition. This walking tour, aided by a headset detailing the stories of the residents, extended the experience of museum going and observing artefacts. One can walk through the factories, hills, old houses, and empty buildings and feel the "exhibition" come alive. In 2014–15, multiple social enterprises and the district government collaborated to promote this idea and framed the neighbourhood as a museum (Seoul Metropolitan Government 2015). They installed sign boards and panels to introduce a street with multiple garment factories as the "Sewing Street Museum of Changsin-dong." A small alley with garment factories was introduced in this street banner, with a written caption that outlines the process of making a garment from scratch. There is no museum building or formal displays: the street itself is decorated and presented as a living museum that shows ongoing factory work.[8]

Whether intended or not, this new cultural commodification of an urban experience sometimes slips into an aesthetic indulgence with spatial and material forms, rendering them remnants of a nostalgic past. Graffiti that says "stay as we remember in our mind" connotes

8 It is notable that, unlike the case of the Chinese History Museum project discussed in Eom's chapter in this volume, the Sewing Street Museum was designed with the agreement of some, if not all, of the residents in the particular alley. However, my analysis focuses on the ambivalent inclusion and exclusion in the symbolic representation of sewing work.

the sentimentality toward this sort of "precious" scene that is vanishing elsewhere in Seoul.[9] Public media and blog postings inspired by the recent revitalization program reproduce this kind of nostalgia and represent the neighbourhood as a place captured in the past for contemplative urban *flaneurs* to enjoy – "an old town giving uncontaminated memories."[10]

The commodification of the place and its spatial experiences extends transnationally, as Changsin-dong has been portrayed in popular Korean dramas as a site where, for example, a young man developed his ambition and desire to succeed in his Dongdaemun fashion business (*Fashion King* [Paesyeon King], 2012); the shabby house of an underprivileged but very independent woman shocked a rich man who fell in love with her (*Secret Garden* [Sikeurit Gadeun], 2010); and a poor young man without much social capital struggles to survive in the corporate world (*Misaeng*, 2014). These dramas' locations in Changsin-dong have recently been introduced in K-pop magazines, guided tours, and walking maps published by the municipal government, along with the newly decorated Sewing Street Museum. Japanese and Chinese tourists comment on their blogs about how they were excited to find the scenes and, for example, felt "shōwa" – the period of Japanese history (1926–89) and a synonym for "the old days" in a general sense.[11] In these walking trajectories in Changsin-dong, the tourists experience and remember the fictive margins of South Korea or their own imagined past.

While these phrases certainly convey more positive images than the dark and depressing ones that the town used to be known for, they have a temporally ambiguous relationship with the goal of bringing the local community and economy together to "regenerate" this space of marginal labour. The nostalgia for an industrial labour and its political organization (Huyssen 2006, 8) is nothing new in many other cities' gentrification processes and city branding for the preferences of consumers from the creative class. As Millington (2013) points out regarding photographers' and tourists' fascination with the emptiness and striking decay of Detroit, the famous industrial

9 A similar tendency also prevails in the revitalization of other areas such as Ihwa-dong, another former garment town.

10 Juhyung Kim, "Changsin-dong: [An Old Town Giving Uncontaminated Memories]," *Yeonhap News*, 16 February 2015, accessed 15 July 2017, http://www.yonhapnews.co.kr/bulletin/2015/02/11/0200000000AKR20150211182900805

11 Henri, "Seoul Travelogue," *Henri Pōtono Burogu* [The Blog of Henri Pōto], 24 August 2015, accessed 1 June 2017, http://ameblo.jp/henry-port/entry-12064678360.html

city in the United States, the nostalgic drive and mournful longing for a Fordist past can be politically ambiguous: they both mourn the city's current state of crisis and industrial ruin while celebrating its picturesque aesthetic. The ambiguity in the commodification process in Changsin-dong also entails the contested aspects of gentrification. On the one hand, the Urban Regeneration Program is praised as an alternative to the gentrification process in which current residents proactively participate in the decision-making process. For example, a television panel on the small communities of Seoul raised an example of Changsin-dong residents' opposition to the plan to control motorbike access in order to protect the work of manufacturing factories.[12] On the other hand, there is far less advancement in the support plan for the network of garment production in the government's Urban Regeneration Program than there is for the beautification process and cultural programs such as the Sewing Museum. The proliferation of nostalgia in the branded place of Changsin-dong is prone to relegate the relevance of garment work to the past as an object with good character, or "cultural content" for the gentrification process, and lead to the displacement of current residents and garment workers.[13]

The public media that has covered the changes in Changsin-dong has sometimes included local residents' responses. Two interviews with Changsin-dong residents, in particular, are notable. Quoted in an online op-ed article, an anonymous resident commented on his experience of visiting the *Made in Changsin-dong* exhibition, pointing out the pitiful (*aejanhan*) representation of Changsin-dong: "The past is important but we asked to highlight the future of the sewing work of Changsin-dong. As we already guessed, though, the exhibition was all about the past and memories. We wished they had also shown the elements of hope in our sewing work" (Cha 2013). Jongim Kim, a sewing worker who has worked for thirty years, also commented on the way the neighbourhood and their work are represented: "Sewing is a great skill and a

12 That is, the residents managed to protect their need for motorbike delivery to sustain their just-in-time production factories, against the move to transform the street for the "walkability" of visitors and passengers. *Dongneui Sasaenghwal* [The Private Life of Neighbourhood] TvN, 27 December 2016, accessed 4 July 2018, http://tvn.tving.com/tvn/vod/view/clip/ea_89460

13 Jiyoon Kim (2016) attributes the complexity in evaluating the gentrification effect to the tension between garment workers and other residents who are not involved in garment work, and the question of the priority between the garment-manufacturing industry and the physical transformation of the neighbourhood.

proper occupation. People pitifully look down on us when they hear that we are working on garment sewing. I hope people just stop talking like 'In miserable condition ... Chun Tae-il struggled so much [to claim labour rights] ...' We are better off this way thanks to him for sure, but we are not that pitiful. We are all just trying to work hard and have an enjoyable life" (Yoon 2013).

The representation of garment work as the more ongoing and current process that these two resident-workers call for would not entirely deny the hardships and existing conditions of their disadvantaged lives. Spurred in opposition to the construction-centred, growth-focused developmental plan, the recent revitalization projects have mobilized the process of change by engaging with the "inside" of local communities and have brought about unprecedented popular attention to the neighbourhood and the people who desire to walk, see, and remember its living spaces. The aspirations and hopes of resident-workers in these interviews seems to emphasize the relevance of their current work for the new future envisioned by the urban project of "regeneration." In a recent study of Changsin-dong's change, Jiyoon Kim (2016, 413) points out that, while the notion of "regeneration" assumes a previous state of decay and stagnation, Changsin-dong was already a vibrant and dynamic neighbourhood before the state-led Urban Regeneration Program. The articulation of people's own sensitivity toward their own living spaces and work will be integral to properly situate the branding of Changsin-dong in the genealogy of urban developmentalism and its evolution.

Reflections on the Unstable Site of Ethnographic Fieldwork and Core Location

This chapter has showed how the word *hyeonjang* emerged in my fieldwork as an ethnographic term used by my interlocutors and in the discourses and policy of the Urban Regeneration Program. Changsin-dong as a "field of garment labour" persisted as an iconic space of disfranchisement in Seoul, where different actors have approached "the field" in their own ways. The connections and disjunctures among these practices highlight the multiple layers and shifting frontlines of this marginalized space, in ways consistent with those discussed by Baik Yeong-seo with respect to the concept of core location (*haeksim hyeonjang*) and by the editors of this volume.

While Changsin-dong is a small neighbourhood in Seoul, its trajectories require a consideration of the sort of "multiple recognition[s] of time-space" (Baik 2013, 62) and of a problematic generated from particularities of the given place. From different angles and perspectives, a particular location can reveal different front lines of privilege, power, and divisions. Changsin-dong, in its different articulations of *hyeonjang*, does not remain merely as a peripheral neighbourhood singularly positioned in the international division of labour or a residual margin of Seoul's uneven development, but emerges as a charged space of dissident relevance, desire, and anticipation. While the neighbourhood was regarded by labour union activists, but also more commonly, as a "gone field" at risk of disappearing, it existed as a continued space of vibrant garment work for workers themselves and other social activists. The active presence of these workers and activists, as well as the collective memory of the vehement activism of the past, opened up a new possibility of this place as a pioneering front line for the new paradigm of urban renewal in the new centuries. Yet, the new phase was not neutral, complicating the meanings and politics of the "field of garment labour" for shifting interests of neoliberal urban governance and city branding, community, and labour advocacy.

The interpretive and performative nature of core location resonates with the anthropological sense of *field* and knowledge production. Baik (2012, 468) calls for an attentive approach to "truthfulness" rather than a positivistic discovery of "truths." Pushing this point further, Sun Ge describes history as not simply a thing that exists outside our bodies, but an undefinable, non-figured situation in which we are all actors who *act upon* history (Baik 2012, 430). The interlocutors of my research, rather than being "figures" in the field, have been active actors who create and transform knowledge of the field through participating in popular discourse, policy documents, and scholarly publications, thereby unsettling the relationship between the field data, scholarly reference, and my own analysis.

While my active fieldwork with Changsin-dong occurred over five years between 2006 and 2011, including preliminary and extended research, I found the most noticeable and visible transformation happened through the urban regeneration project that was enacted after I returned to California. The project was conducted and new social enterprises proliferated swiftly. The central issue during my fieldwork was to examine the first disjuncture that I discussed earlier in this chapter – that is, the popular ideas and developmental frame that assumes that garment labour is all gone or temporally obsolete and the subjective claims of garment workers and their persistent investment and attachment to their work. However, the presence of hidden and marginal garment labour started to be widely promoted by the recent urban regeneration project and social enterprises in Changsin-dong. With the emerging excitement for regenerating the community and popular interest in a nostalgic urban experience, calling for attention was not an issue anymore; rather, the question was, what kind of attention it received.

As I was collecting newspaper and archival data as follow-up research, I also recognized quite a few individuals whom I had encountered or interviewed during my earlier research. Under new conditions where they gained much greater attention, and in the changing environment, people including my former interlocutors were offering their own ideas to media interviewers and other researchers. Some of these accounts emphasized the location of Changsin-dong in the recollection of their work and life. Some contained ideas that resonated with the content of interviews that I had previously considered too peculiar and chose not to include in my dissertation. While I have already written about the attachment and passion that people had for garment work despite their own marginality, the newly available narratives and their slightly different tones have impacted my subsequent writing and interpretation of "marginality" to varying degrees. While my representation of workers' experience emphasized the continuity and persistence of their presence against the evolutionary narrative of the transition in the mode of material production, the new circumstances have compelled me to rewrite my analysis of their belonging to the projected future by the urban-planning projects.

The complexity of the field site that I needed to examine stemmed not merely from the fact that there happened to be more historical contingencies in Changsin-dong; rather, I had to go back to what I had considered "data" and memories to find out where I could discover the ambiguous and murky spots through which I could take a different analytic perspective toward the area's seemingly marginal and disfranchised labourers. The genealogies of the labour movement, garment works, and urban policies, as well as of placed-based activism and community building inform each other to reframe the temporality of "the field of garment labour," rather than representing a fixed phase in its history. The process of this reflection was more than "updating" the field – rather, it was locating my analysis and reference point within the shifting relevance of my "field." That is, I had to consider whose perspective and practices would most actively construct the site of my empirical study and reveal the most meaningful relevance to the sense of work and history of my interlocutors.

The demand of relevance challenges the belatedness and slowness of ethnographic work (Marcus 2013) and destabilizes the "field," as is much discussed in anthropological explorations. Akhil Gupta and

James Ferguson (1997) have challenged the temporal and spatial distance between the location of a field and the location of one's academic home, claiming that it is no longer tenable to maintain a before-and-after idea of the fieldwork and writing processes. Simon Coleman and Peter Collins (2006) have captured "location" as a *verb* and argue that "locating" is a disciplinary practice within an ethnographic field – including a strategic "dislocation" by micro-negotiations in the field to see the dynamic complexities of the site. Different meanings of *hyeonjang* suggest that the field is "an emergent form, and a localized expression of a distinct institutional apparatus" (Coleman and Collins 2006, 5) as well as a milieu of "temporal dwelling," given that not only the object of study but also what we take to constitute a site might already be in the process of transformation (Dalsgaard and Nielsen 2013, 9).

While discussions of the "field" of empirical knowledge as particular, changeable, and lived space is not entirely new, thinking the "field" (*hyeonjang*) and "core location" (*haeksim hyeonjang*) together with the common word *hyeonjang*[14] reveals the intertwined conceptual, physical, social, and ethnographic spaces of labour. In each moment and modality of the "field of labour," in the case of the physical location of Changsin-dong, actors not only create a different social scene that others observe or document, but they also create varying perspectives and knowledge through which they frame their own politics and praxis. In this sense, *hyeonjang* does not remain merely at a conceptual level but works as a lived space for multiple actors whose "experience" and "ideas" in and of the place constantly destabilize the field of empirical study. Labour activists, social entrepreneurs, and urban planners often are not merely research subjects. Rather, in many cases, they operate within academic institutions and produce reports and research papers, and labourers produce personal memoirs and public commentaries of their experiences, a process through which they also *act upon* the field and locate themselves.

While *hyeonjang* as an ethnographic term can be translated into different spatial referents, as a field of praxis and political mobilization, a site of work and life, a lively scene of marginal labour, and a fieldwork site, the first morpheme, *hyeon*, commonly conveys a material and affective "present-ness" – fundamentally temporal – that is immediate and emerging. The discussion above adds a more pronounced will and need to imagine and create connections between workers, activists, and scholars

14 Both ideas are termed as *hyeonjang* in Korean, while *jiyeok* or *jijeom* would be the more frequent term for "location" in other contexts.

than the usual anthropological framing of location. It reveals a different sphere of social engagement and dissident temporal orientations that create the "disjuncture." The disjuncture was the most noticeable of the temporal orientations of these different registers of space – ranging from the praxis coming to an end to the new future emerging from a nostalgia for ruins. The figures of "the past" seek a relevance in the future of the newly branded neighbourhood, shifting the front line of their own marginality, impacting the way I close and open the time span of my own analysis of research. A particular history in which labour matters in a city cannot be a process of linear transition from the industrial to post-industrial, or from central to peripheral. Rather, the case of Changsin-dong reveals that we can understand only a particular way in which a specific kind of labour *matters,* through intersecting physical, epistemological, and social spaces where different social actors understand, act upon, and construct the meanings of labour.

REFERENCES

Baik, Young-seo. 2012. "Sinjayujuuisidae Hangmunui Somyeonggwa Sahoeinmunhak: Ssun-geowaui Daedam" [The Intellectual Call in Neoliberal Era and Social Humanities: Dialogue with Sun Ge]. *Dongbangjihak* [Eastern Knowledge] 159: 421–71.

Baik, Young-seo. 2013. *Haeksimhyeonjang-eseo Dong-asiareul Dasi Mutda: Gongsaengsahoereul Wihan Silcheon-gwaje* [Rethinking East Asia in Core Locations: A Task for a Co-Habitation]. Paju, KR: Changbi.

Berg, Nate. 2012. "Seoul Ends Failed 'New Towns' Project." *Atlantic City Lab,* 6 February. Accessed 17 July 2017. https://www.citylab.com/equity/2012/02/seoul-ends-failed-new-towns-project/1149/

Bonacich, Edna, and Richard Appelbaum. 2000. *Behind the Label: Inequality in the Los Angeles Apparel Industry.* Berkeley: University of California Press.

Byeon, Chang Heum. 2013. "Nyutaunsaeobui Jeongchaeksilpae Aksunhwan Gujo Bunseok" [Analysis of Vicious Cycle of Policy Failure: Case Study on the New Town Project in Seoul]. *Gonggangwa Sahoe* [Space and Society] 44: 85–128.

Cha, Seho. 2013. "'Meideu In Changsin-dong' Yugam" [Discontent with Made in Changsin-dong]. *Hanguk Gyeongje* [Korean Economy], 29 June. Accessed 17 July 2017. http://snacker.hankyung.com/life/1272

Cheng, Tun-jen. 1990. "Political Regimes and Development Strategies: South Korea and Taiwan." In *Manufacturing Miracles: Paths of Industrialization in Latin America and East Asia,* 139–78. Princeton, NJ: Princeton University Press.

Cho, Hee Yeon. 1988. "Student Movement in 1980s and Its Unfolding Process." *Sahoe Bipyeong* [Social Critiques] 1: 124–50.

Cho, Mun Young, and Seung Chul Lee. 2017. "Sahoeui Wigiwa 'Sahoejeogin Geos'ui Beomnam: Han-gukgwa Junggugui Sahoegeonseol Peurojekteu Gwanhan Sogo" [The Crisis of "Society" and the Explosion of "the Social": Social Construction Projects in South Korea and China]. *Gyeongjewa Sahoe* [Economy and Society] 113: 100–46.

Cho, Soon Kyoung. 1985. "The Labor Process and Capital Mobility: The Limits of the New International Division of Labor." *Politics and Society* 14 (2): 185–222.

Chu, Nellie. 2016. "The Emergence of 'Craft' and Migrant Entrepreneurship along the Global Commodity Chains for Fast Fashion in Southern China." *Journal of Modern Craft* 9 (2): 193–213.

Chun, Jennifer Jihye. 2011. *Organizing at the Margins: The Symbolic Politics of Labor in South Korea and the United States.* Ithaca, NY: Cornell University Press.

Chun, Soonok. 2017. *They Are Not Machines: Korean Women Workers and Their Fight for Democratic Trade Unionism in the 1970s.* London: Routledge.

Coleman, Simon, and Peter Collins. 2006. "Introduction: 'Being … Where?' Performing Fields on Shifting Grounds." In *Locating the Field: Space, Place and Context in Anthropology.* 1–21. Oxford: Berg.

Dalsgaard, Steffen, and Morten Nielsen. 2013. "Introduction: Time and the Field." *Social Analysis* 57 (1): 1–19.

Gupta, Akhil, and James Ferguson. 1997. "Discipline and Practice: 'The Field' as Site, Method, and Location in Anthropology." In *Anthropological Locations: Boundaries and Grounds of a Field Sciene,* edited by Akhil Gupta and Games Ferguson, 1–47. Berkeley: University of California Press.Han, Na. 2017. "Historical Reconfiguration of Changsin-Dong's Community Care: Sewing Factories, Local Movement, Community Project, and Urban Regeneration." MA thesis, Yonsei University.

Huyssen, Andreas. 2006. "Nostalgia for Ruins." *Grey Room* 23: 6–21.

Kim, Jiyoon. 2016. "Changsin-dong: Geullobeol Dosimandeulgiwa Dosijaesaeng Sai" [Changsindong: Between Global City Making and Urban Regeneration]. In *Seoul, Jenteuripikeisyeoneul Malhada* [Seoul: Speaking Gentrification], edited by Hyeonjoon Shin and Giwoong Lee, 360–416. Seoul: Pureunsup.

Kim, Seung-keyong. 1997. *Class Struggle or Family Struggle? The Lives of Womenn Factory Workers in South Korea.* Cambridge: Cambridge University Press.

Koo, Hagen. 2000. "The Dilemmas of Empowered Labor in Korea." *Asian Survey* 40 (2): 227–50.

Korea Social Enterprise Promotion Agency. 2016. "Sahoejeokgieob Injeung Hyeonhwang" [The Current Certification of Social Enterprise]. 29 February. Accessed 15 March 2016. http://www.socialenterprise.or.kr/notice/docu_books.do?page=1&cmd=&search_word=&board_code=BO04&category_id=CA06&seq_no=226623&com_certifi_num=&selectyear=&mode=view

Marcus, George. 2013. "Afterword: Ethnography between the Virtue of Patience and the Anxiety of Belatedness Once Coevalness is Embraced. *Social Analysis* 57 (1): 143–55.Millington, Nave. 2013. "Post-Industrial Imaginaries: Nature, Representation and Ruin in Detroit, Michigan." *International Journal of Urban and Regional Research* 37 (1): 279–96.

Mills, Mary Beth. 2003. "Gender and Inequality in the Global Labor Force." *Annual Review of Anthropology* 32 (1): 41–62.

Park, Chanim. 2008. "Sahoejeok Gieob Changchul Mit Yukseong-eul Wihan Gwaje" [Tasks for Creating and Fostering Social Enterprises]. *Wolgan Nodong Ribyu* [Monthly Labor Review] 43: 31–48.

Park, Seo Young. 2012. "Stitching the Fabric of Family." *Journal of Korean Studies* 17 (2): 383–406.

Park, Seo Young. forthcoming. *The Problem of Speed: Working in the 24-hour City of Seoul.* Ithaca, NY: Cornell University Press.

Seoul Metropolitan Government. 2015. "Changsindong-e Bongjebangmulgwan, Bongjegeori Joseong" [Building Sewing Museum and Sewing Street in Changsin-dong]. *Seoul Information Communication Plaza,* 16 October. Accessed 3 July 2018. https://opengov.seoul.go.kr/mediahub/6355532

Seoul Metropolitan Government. 2016. "'Seoulhyeong Dosijaesaeng 1Ho' Changsinsung-in Saeob Bon-gyeokhwa" ["The First Seoul-type Urban Regeneration" Changsin-Sungin Project Gets into Stride]. *Seoul Information Communication Plaza,* 21 July. Accessed 5 July 2018. https://opengov.seoul.go.kr/mediahub/9130136

Shin, Hyun Bang. 2017. "Introduction." *Anti Jenteuripikeisyeon: Mueoseul Hal Geosin-ga* [Anti-Gentrification: What We Should Do]. Seoul: Dongnyeok.

Shin, Hyun Bang, and Soo-Hyun Kim. 2016. "The Developmental State, Speculative Urbanisation and the Politics of Displacement in Gentrifying Seoul." *Urban Studies* 53 (3): 540–59.

Sun, Ge. 2003. *Asiaraneun Sayugonggan* [Asia as a Space for Thinking]. Translated by Junpil Ryu. Seoul: Changbi.

Yoo, Kyeong Soon. 2013. "Haksaeng-undonggadeurui Nodong-undong Chamyeoyangsanggwa Yeonghyang: 1970 Nyeondaereul Jungsimeuro"

[Labor Participation Patterns and Influence of the Student Activists: The 1970s Story]. *Gieokgwa Jeonmang* [Memory and Prospect] 29: 52–97.

Yoon, Hyungjoong. 2013. "Jeon Tae-ildo Uricheoreom Jaeminage Salgo Sipeotgetjyo" [Jeon Tae Il also Would Have Wanted to Live a Fun Life]. *Hankyoreh*, 9 August. Accessed 1 July 2018. http://www.hani.co.kr/arti/society/labor/598987.html

Afterword

JESOOK SONG AND LAAM HAE

Readers of this volume might wonder how we, the editors and contributors, came to conduct this thematic and paradigmatic inquiry related to "Asia as method" and "core location." Indeed, it has been the product of a long journey through various platforms. Our collective point of departure was a shared interest in interrogating the paradigms through which urban South Korea has typically been studied. Many aspects of the project, however, were not initially foreseen. For one, none of us expected the lengthy timeframe it would ultimately require to complete the volume when we first met on a panel on urban Korea at the annual meeting for the Association for Asian Studies (AAS) in Chicago in March 2015, which was the beginning point of our discussion regarding developing a project together.

In a way, our thematic attention to urban Korea has remained intact – although it was expanded to encompass certain aspects of rural and transnational Korea (see the chapters by Cho, Eom, Jeong, and Oh, in particular) – attempting to challenge the boundary of urban and rural, and national and transnational, dichotomies. Yet, we did not anticipate the changes in our framework from urban development to Asia as method and core location, or the format of publication from a special issue in a journal to an edited volume. This has been quite a dialectical, experimental, and collaborative process, though not one without its uncertainties and challenges. This afterword affords us the opportunity to reflect upon and share the journey of our collaborative efforts. We undertake this not necessarily as an account of the unusual trajectory that we took to produce this volume, but as a testament to the potential of open-ended intellectual endeavours that cherish contingent processes, rigorous dialogues, and a commitment to collaboration. Toward these ends, we met as a group on five conference panels and via workshops, punctuated by three Skype conversations for discussions of readings and writings.

Youjoeng Oh, a contributor to this volume, organized double panels on the subject of urban developmentalism in South Korea at the 2015 AAS meeting in Chicago, inviting as panelists scholars whose research was broadly connected to this theme. Many of the contributors to this volume presented papers as part of these panels. Oh also convened a post-conference meeting to explore the possibility of publishing the presented papers as a special issue of a journal. There was nearly unanimous support for this idea, arising from the excitement and appreciation for the value of the presented papers, which made a range of contributions to interdisciplinary understandings of Korean urban development and urban developmentalism in general. Most of the participants were not particularly experienced in the publishing process, however: the majority were junior scholars who had recently secured tenure-track positions or who had just finished their doctoral degrees. In light of this, the editorial responsibility was left to the two people who were in relatively senior positions at that time – that is, Jesook Song and Laam Hae. These two scholars began exploring different journals in the fields of area studies, development studies, and urban studies that might be willing to host our collection as a special issue. Given the fact that the academic careers of many contributors were facing "publish or perish" pressures, speedy publication of their work within a couple of years was the goal, especially in peer-reviewed journals. Despite our efforts, however, our inquiries to these journals did not yield any success.

In order to strengthen our manuscript, Jesook Song and Laam Hae organized a two-day workshop with the other contributors in the fall of 2015, co-sponsored by the York Centre for Asian Research (YCAR) at York University, Toronto, and the Centre for the Study of Korea (CSK) at the University of Toronto. For this workshop, we invited esteemed scholars in geography (Jim Glassman – the University of British Columbia), critical development studies (Katharine Rankin – the University of Toronto) and Asian urban studies (Bae-Gyoon Park – Seoul National University). We received insightful suggestions from them that were helpful in revising and refining our manuscripts for publication.

Based on the strengthened manuscripts that emerged from the first workshop, we submitted our proposal to a wider set of journals. Unfortunately, these efforts were also made in vain. This was a valuable learning process for us, however, as the rejections by the journals illuminated several things to us. First, the limited frequency of the publication of special issues and consequent backlogs of special issues to come out in the next few years was a common basis for many top-tier journals to decline taking on new special issue proposals. In particular, there has been increasing competition among Korean studies scholars to publish journal articles, as a result of the current incentive and funding structures, especially for scholars based in Korea. This has inflated the number of requests for special issues in various journals, which has led most to refuse requests for any special issue themes related to Korean studies. Second, journal editors seemed to think that a special issue that focuses primarily on a specific country would not be appealing to readers who have increasingly preferred transnationalist perspectives. The assumption, it seems, was that research on a particular country (i.e., South Korea) was incapable of engaging in transnational analysis. Third, some journals were concerned about the disciplinary fit of our project for their audience, as our manuscripts engaged with both social sciences and humanities perspectives and methodologies.

By the time we were preparing for a panel at the Annual Meeting of the Association of American Geographers (AAG) in San Francisco in April 2016, the editors of and contributors to our project were puzzled by the repeated rejections by the journals and the rationales given for doing so. Our puzzlement and predicament was twofold, related to the form of publication and the value of our research. Regarding the form of knowledge production and distribution, we began to question whether a special issue in a journal was necessarily the best option for us. Second, we confronted the difficulty of assuring the significance of knowledge grounded in a particular location, especially one in the non-West, because of its supposed limitations in appealing to conceptions of universality and certain understandings of transnationality. Thus, we began discussing how to reframe our work in terms of engaging in conceptual dialogues, rather than as a collection of case studies of a particular location. We believed that our research problematic was neither parochial nor irrelevant to a general or cross-regional audience, so we felt it imperative to frame our problematic more clearly in transnational and relational terms. Additionally, we also felt it necessary to highlight our interest in frameworks that challenge Euro-American epistemologies and anglophone hegemony in the field of knowledge production.

Following the 2016 AAG meeting, then, we started to examine the tradition of "Asia as method" as a tool to rethink our research subject and our interest in studying different locations in and of the Koreas. This approach also gave us a chance to concretize our take on epistemological colonization, and provided us with an analytic lens through which to interrogate the validity of postcolonial schools of thought. In an effort to situate critical approaches to development in understanding South Korean capitalism (on which most of our manuscripts are foregrounded) along with the humanistic literature on Asia as method, we found another venue of exploration that was opened to us at a workshop at Seoul National University, co-organized by Bae-Gyoon Park and Jesook Song in July 2016. We read and discussed various leftist state theories (Keynesianism, developmentalism, and neoliberalism) in conjunction with the Asia as method literature for the workshop. While it was a productive and exciting cross-fertilizing moment for us, it also confirmed the well-entrenched disconnect within contemporary academia between postcolonial humanities, on the one hand, and the metatheory of development studies and social sciences on the other. Ironically, the fruitfulness of our explorations came from a meeting after the workshop. In the debrief meeting for the workshop, the members stumbled fortuitously upon a notion that would provide an original platform to reveal our manuscript's significance in terms of both content and approach. It was Mun Young Cho's erudite understanding of the multiple intellectual lineages within Asia as method that brought Baik Yeong-seo's concept of "core location" to our attention. Since that time, we have conducted three Skype meetings among the writers of this volume, who are currently spread across three continents, by reading and discussing together the genealogies of Asia as method, including Baik's literature in Korean and English on core location.

Through intense discussions and debates about the idea of core location, we gradually reorganized our conceptual stance by synthesizing works on postcolonialism, Asia as method, core location, Marxist area studies, and leftist urban studies. In order to revise our manuscript in accordance with this major turn, we organized another conference panel at the 2017 AAS meeting in Toronto (organized by Hyeseon Jeong), as well as at a publication workshop at the University of Toronto in October 2017 (hosted by Jesook Song). At these meetings, we continued to exchange rigorous feedback on each contributor's paper. We also benefited from the critical comments of Hyun Bang Shin, a leading geographer and urban studies scholar who explores planetary processes of gentrification in dialogue with postcolonial urban scholarship.

During this process, we decided to seek publication of our manuscripts in the form of an edited volume, instead of a special issue. This decision was the result of a long-term collective commitment to explore and seek an original way through which to understand urban development in South Korea without losing sight of the significance of the knowledge of it in relation to the problematic of decolonizing knowledge production and the broader matrix of political economy and resistance politics. Moreover, without the integrity, patience, and solidarity of the members in less-secure jobs who nonetheless took the initiative of this collective project and stood by every turn in its development, we could not have come this far as a collective. This was a tremendously valuable experience of intellectual collaboration, one that cannot be taken for granted in the current hyper-competitive academic world.

Through our journey of knowledge production, we learned the importance of open-ended questioning and critical intellectual engagement, realizing that the value of our research content and material is not necessarily self-evident when viewed through the epistemological hegemony of Western-cum-anglophone knowledge circulation. Yet, the challenges we faced on the way, and will doubtless face again in the future, catalyzed us to explore unfamiliar paradigms together and build our thoughts from our tenacious collaboration. This kind of open-ended and long-engaging dialogue over ideas generated decades ago resembles the praxis that Sun Ge referred to as "building ideas" – that is, not merely deciphering ideas, but rather making sense of the insights in our own spatial-temporal problematics. This volume was, therefore, an experiment in this sort of praxis, similar to the ways in which Sun Ge excavated Takeuchi Yoshimi's ideas behind Asia as method in her geo-historical juncture, or to how Takeuchi Yoshimi reinterpreted Lu Xin's ideas of self-disavowal for his time and place in Japan.

Contributors

Mun Young Cho is an associate professor in the Department of Cultural Anthropology at Yonsei University, South Korea. Her research interests include poverty, labour, development, and youth in China and South Korea. Based on fieldwork in Harbin, Cho's book *The Specter of "The People": Urban Poverty in Northeast China* (Cornell University Press, 2013) examines China's poverty management as the nation's fast-growing market economy brings about deeper impoverishments of one-time socialist workers. She has also translated into Korean James Ferguson's *Give a Man a Fish: Reflections on the New Politics of Distribution*. From social work projects targeting Foxconn workers-cum-volunteers in Shenzhen to social innovation practices among precarious youth in Seoul, Cho currently charts the changing topography of "the social" in China and South Korea, focusing on what it reveals about the experiential and existential conditions of poverty. Regarding this research, she has a number of articles in journals and edited volumes in both Korean and English.

Sujin Eom is a postdoctoral fellow in the Department of Geography at Dartmouth College, New Hampshire. She received a PhD in architecture from the University of California, Berkeley, with a Designated Emphasis in Global Metropolitan Studies. She is currently completing her first book, which examines the transpacific migration of racial and spatial categories, through the lens of Chinatowns. Her research interests include migration and diaspora, critical race theory, science and technology studies, and postcolonial urban theory. Her research has appeared in *positions: asia critique*, *Planning Perspectives*, and the *Traditional Dwellings and Settlements Review*. Her second book-length project investigates the global movement of urban policies, ideas, and forms during the Cold War, with an emphasis on urban infrastructure built across Asian cities.

Laam Hae is an associate professor in the Department of Politics at York University. Her research interests span the subjects of urban political economy and redevelopment, socialist feminist politics, legal geography, and urban social movements in the context of South Korea and North America. Her book *The Gentrification of Nightlife and the Right to the City: Regulating Spaces of Social Dancing in New York* (Routledge, 2012) explores how gentrification and neoliberal urbanization in New York City transformed legal, cultural, and justice geographies of the city, and makes a normative argument for the restoration of the right to the city. She has also written on the intermediating roles of critical urbanists in the mobilization of place marketing strategies in Korea. She is currently researching childcare co-op movements in Korea, engaging with socialist feminist theories of activist social reproduction.

Hyeseon Jeong is a political geographer based at the University of Newcastle in Australia. She has taught geography and urban affairs at Ohio State University and Wright State University, and conducted research for the government of South Korea to make its international development assistance projects gender-responsive. Her research interests include international aid, community development, transnational migration, gender, and territorialization. Her writings have been published in *Geoforum, the Journal of International Development Cooperation,* and the *Handbook of the Changing World Language Map.*

Youjeong Oh is an assistant professor at the University of Texas at Austin. Her recent book, *Pop City: Korean Popular Culture and the Selling of Place,* examines the political economy of place promotion through the lens of Korean television dramas and K-pop music. She demonstrates how the speculative, image-based, and consumer-exploitive nature of popular culture shapes the commodification of urban space and ultimately argues that pop culture–mediated place promotion entails the domination of urban space by capital in increasingly sophisticated and fetishized ways. Her second project investigates the Jeju Free International City development project, which connects Jeju's histories of dispossession and marginalization to current neoliberal and speculative forms of developmentalism.

Seo Young Park is an associate professor of anthropology at Scripps College, in California, with research and teaching interests in labour, gender, commodity production and consumption, and urban East Asia. Her first book, *Working in the 24-Hour City: The Problem of Speed in Seoul's Dongdaemun Market* (Cornell University Press, forthcoming), illustrates the lived experiences of South Korean garment workers and analyses how their intimacy, affect, and embodied attachment to work shape the temporality of labour and the urban fabric. She is currently working on an ethnographic project that explores the ways in which everyday toxicity, anxiety, and bodily experience of indoor air quality become significant public knowledge in the urban processes and environmental politics of Seoul.

Jesook Song is a professor in the Department of Anthropology at the University of Toronto. Her teaching focuses on the anthropology of neoliberalism, gift money finance, transnational Korea, representations of intellectuals, and anthropology of activism and social justice. Her research interests include finance, welfare, labour, education, gender, sexuality and reconciliation. Her first book, *South Koreans in the Debt Crisis* (Duke University Press, 2009), deals with homelessness and youth unemployment during the Asian financial crisis in the late 1990s to early in the millennium. Her second book, *Living on Your Own* (SUNY Press, 2014), is about single households and informal financial markets from the perspective of single women's struggles in South Korea. Both of her single-authored books were translated into Korean in 2016. She has also published an edited volume, *New Millennium South Korea: Neoliberal Capitalism and Transnational Movements* (Routledge, 2010), co-edited *Korea through Ethnography*, a special issue of the *Journal of Korean Studies* (with Nancy Abelmann, November 2012), and has a number of articles in journals and edited volumes.

www.ingramcontent.com/pod-product-compliance
Lightning Source LLC
LaVergne TN
LVHW090159080826
844660LV00013B/873/J

* 9 7 8 1 4 8 7 5 0 3 3 5 2 *